Hermann Mascher
**HPLC Methods for Clinical
Pharmaceutical Analysis**

Related Titles

Xu, Q. A., Madden, T. L.

Analytical Methods for Therapeutic Drug Monitoring and Toxicology

502 pages · Hardcover

2011
ISBN: 978-0-470-45561-6

Meyer, V.

Practical High-Performance Liquid Chromatography

426 pages · Softcover

2010
ISBN: 978-0-470-68217-3

Snyder, L. R., Kirkland, J. J., Dolan, J. W.

Introduction to Modern Liquid Chromatography

960 pages · Hardcover

2010
ISBN: 978-0-470-16754-0

Wixom, R. L., Gehrke, C. W.

Chromatography - A Science of Discovery

410 pages · Hardcover

2010
ISBN: 978-0-470-28345-5

Kazakevich, Y. V., LoBrutto, R.

HPLC for Pharmaceutical Scientists

1136 pages · Hardcover

2007
ISBN: 978-0-471-68162-5

Hermann Mascher

HPLC Methods for Clinical
Pharmaceutical Analysis

A User's Guide

WILEY-VCH

WILEY-VCH Verlag GmbH & Co. KGaA

The Author

Prof. Hermann Mascher
pharm-analyt.Labor GmbH
Ferdinand-Pichlergasse 2
2500 Baden
Österreich

Library of Congress Card No.:
applied for

British Library Cataloguing-in-Publication Data
A catalogue record for this book is available from the British Library.

Bibliographic information published by the Deutsche Nationalbibliothek
The Deutsche Nationalbibliothek lists this publication in the Deutsche Nationalbibliografie; detailed bibliographic data are available on the Internet at [http://dnb.d-nb.de].

© 2012 Wiley-VCH Verlag & Co. KGaA, Boschstr. 12, 69469 Weinheim, Germany

Typesetting Primustype Robert Hurler GmbH, Notzingen
Printing and Binding Markono Print Media Pte Ltd, Singapore
Cover Design Grafik-Design Schulz, Fußgönheim

Printed in Singapore
Printed on acid-free paper

Print ISBN: 978-3-527-33129-1

Content

Preface

For more than 35 years I have been doing bioanalytical research. I have again and again searched for and wanted a reference book comprising the most important information on substances to be analyzed: on the structure of molecules, on the UV spectrum, on the Cmax value after oral application, on the pKa value, on the protein binding and the elimination half-life as well as on published bioanalytical methods and their limits of determination. With the publication of this book I myself fulfil my wish, but I do know that there are many analysts doing bioanalytical research who have the same wish.

The aim of this book is to show a specific procedure in clinical analysis. It would be quite the same with eating. There are different ways of putting food in your mouth and so there are also various possibilities of approaching an analytical problem.

While the Chinese take chopsticks as a matter of course and use them, I am used to spoons, forks and knives as means to eat and this procedure has also proved itself. In exactly the same way it proved to be efficient in my practical bioanalytical work to start the method development by thinking about detection, in the end fixing the determination limit to detection. In this book you will get to know many examples referring to this subject. Carrying out research on various analytical questions for a couple of decades and discussing with colleagues at work and at congressional and application meetings I got a general way of thinking as well as a profound knowledge of specific individual problems. This book deals with my experience and thus I only seldom consulted other literature or articles offering an overview of the subject. It does not offer a current overview of specific subjects, but it gives an overall view of possibilities, which have proved to be very good, as well as including mistakes we happened to make. During a couple of decades naturally many discussions with my colleagues took place and have had much influence on this book. But in the end, this book deals with my perspective on possibilities and potential errors within the subject of clinical HPLC. If any information in this book turns out to be wrong, I myself am the only one to be blamed.

So, I hope that you learn a lot and have many sudden insights, but also that you may smile now and then while reading this book as a whole or partially.

Did the information in this book help you to deepen your knowledge? Or do you still have questions? Do you want to make an appointment for lectures or trainings? I will be pleased to receive your feedback! At www.pharm-analyt.at you are welcome to get in contact with me.

Readers' comments on the German edition

"Klinische Analytik mit HPLC - Ein Ratgeber für die Praxis" (ISBN 978-3-527-32751-5)

„I found the book to be 100 % useful for my day-to-day work. The author offers a perfect balance of analytical data and expert advice."
Franz Kricek, NBS-C BioScience

"The book really fills a gap. It is written in a clear and didactic style that makes the complex subject accessible even to non-specialists."
Maria Theresia Kaltwasser, Berlin Chemie – Menarini Group

"The practitioner will be pleased to discover a host of eminently useful tips and tricks that are otherwise hard to find."
Karl Zech, Nycomed

"A rare example of the scientific literature that combines supreme competence and expert knowledge with style and – occasionally – humor."
Andreas Grassauer, Marinomed

"It will become the primary reference for analytical questions in our laboratory."
Herbert Korall, Center for Metabolic Diagnosis Reutlingen

Acknowledgements

In the course of my "life as an analyst" a number of people had a positive influence on me. My first and crucial teacher of analytics was Dr. Franz Lorenz of Biochemie Kundl (nowadays Sandoz), who always forced me to look behind the "analytical curtain". It is a great joy for me that my eldest son, Dr. Daniel Mascher, follows in my "analytical footsteps" and he has already solved many analytical questions in a brilliant manner especially with HPLC-MS/MS. I give thanks to my friend and comrade-in-arms for more than 20 years, Mag. Christian Kikuta, because he managed the laboratory (with regard to GLP/GMP) and solved other analytical problems and thus made it possible for me to have more time for creative work. I also thank my beloved wife Christiane in connection with this book and its contents because she was very open to my inquiring mind and thus made many creative thoughts and ways possible.

I am also grateful for friends and acquaintances who have given me pieces of advice and access to analytical documents. I also thank my younger son, Mr. Klaus Mascher, and Mrs. Maria Pech for helping with the English translation. I also thank the Editio Cantor Verlag (Mr. Andreas Gerth) for making the most of the UV spectra available to me for a reasonable price, which give to the book a strong foundation in reality. Mr. Frank Weinreich, Waltraud Wüst and Mrs. Bernadette Gmeiner supported the idea of this book and its development and helped to make the approach of this book a practical one.

Baden, April 2011   *Hermann Mascher*

1
Introduction

When you as an analyst are facing the challenge of determining a certain sub-stance in plasma, urine or tissue – it may be a drug, a metabolite of a drug or a substance belonging to an organism (= endogenous substance), then totally dif-ferent approaches are possible. You could start with the question: "What has been published on this subject so far?" It can easily happen that a multitude of possibili-ties overwhelms you, especially with drugs that are frequently prescribed. As an alternative, you can contact friends and acquaintances among your colleagues, who have hopefully already had experience with that substance in order to get some hints. A third possibility would be an intellectual time out, which, however, often ends with a chaotic gathering of material and thus possibly creates more questions than it yields answers.

Have you not always looked for a tool to help you with such questions, with which you can approach a solution in due time and that quickly informs you whether your own instruments suffice to reach your aim? In this book you shall get to know an adequate approach. Parts of this concept are familiar to every analyst who is in clinical/pharmaceutical research. However, the radical nature of this approach that has proved its worth a thousand times with us and our work, makes it a means that you may like to apply again and again. The following three chapters each introduce a specific question and develop a solution for it. Please, apply at first your method to develop a solution and then compare, if the approach of this book is familiar to you and you are used to applying it anyway, or you perhaps get to know new tools in this book, which may help you to solve the questions quickly and well.

Plasma (serum) plays an important role in clinical practice or with bioanalytical tests performed by pharmaceutical laboratories. Occasionally, urine checks are also of great importance: when you observe and check the characteristics of an illness, when you supervise the therapy of an illness or when you prepare a mass balance of drugs (e. g., how much of an oral applied medication you find in the excrements of the patient). Whether the substance under study is the parent drug or is one or more of the numerous metabolites, is not important right now.

HPLC Methods for Clinical Pharmaceutical Analysis, First Edition. Hermann Mascher
© 2012 Wiley-VCH Verlag GmbH & Co. KGaA. Published 2012 by Wiley-VCH Verlag GmbH & Co. KGaA.

1.1
First Question: Determination of Ibuprofen in Plasma

A colleague or a superior asks you to determine the concentration of an NSAR (nonsteroidal antirheumatic) in human plasma. Possibly the superior being a medical person says something like: "You have already analyzed substances in this field of therapy, thus you could determine this substance – couldn't you?" If you ask for further information, you possibly do not even get the name of the substance, but merely the name of the drug – the trade name – and perhaps the dosage. The superior, a clinical employee in our case, has administered Brufen and therefore needs – that is what you can see after you have had a short glance at the list of drugs – the determination of ibuprofen in human plasma. Now – how do you usually approach such a question? Do you look for a reference in technical literature? Do you call colleagues? Do you have a glance at the structural formula?

According to our experience we recommend to answer the following two questions first:

- Which determination limit is needed?
- Which kind of detector is at your disposal?

It is easy to answer the second question. For this example we suppose you have an HPLC instrument, a UV detector with variable wavelength and a fluorescence detector at your disposal. However, the question about the determination limit is much more difficult. What question does the superior want to solve? Does he need a check on the patient's compliance with the therapy? Is he concerned about too high plasma levels due to certain side effects? Or does he merely like to get to know minimal plasma concentrations before the next medication dosage ($C_{ss\ min}$)? For a maximum level at a usual medication dosage you find in the literature (e. g., "Martindale" is a good reference book[1]; or information given by the manufacturers of the drug) an approximate value of 20–50 μg/ml (see Appendix "Ibuprofen"). Also, details on elimination half-life are often important with such questions because they tell you when the clinical employee should draw the blood samples in order to answer his questions. Besides, the elimination half-life tells the analyst the plasma level he may expect, if – according to standard rules – blood samples are drawn approx. 5 h after the drug has been administered orally.

Back to the question: The superior in the given example is merely interested in the maximum level, because of the side effects. Thus, he must know that he has to carry out blood drawing after approx. 2 h and not right after oral administration or after 5 h even.

The analyst now knows that he has got to develop a safe proof within 10–60 μg/ml plasma. Actually, it looks like it would be simple: 10–60 ppm, whereas often proof is requested within the range of 10–60 ppb (10–60 ng/ml). At this point of development I recommend that you have a glance at the structural formula and at – as only UV and fluorescence detection are at you disposal – solubility as well as at the UV spectrum.

---

1   See list of books: Martindale

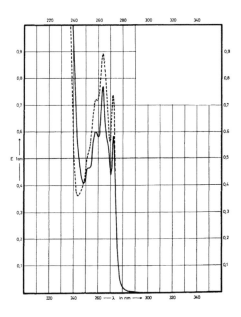

MW: 206.3

Table 1.1: Ibuprofen – Details to the UV-Spectrum

solvent symbol	methanol ———	water — · — · —	0.1M HCl — — — —	0.1M NaOH · · · · · ·
absorption maximum	272 nm 264 nm 258 nm			272 nm 264 nm 258 nm
$E_{1cm}^{1\%}$	11.2 14.5 11.3			15.4 18.4 15.0
ε	230 300 233			320 380 310

- *Structure:* Ibuprofen tends to be a lipophilic substance, but it contains a carboxy function.
- Solubility: As a nondissociated substance in the acidic pH range it is lipophilic, as a salt (pH >6), however, presumably very well soluble in water.
- UV spectrum: Only below 230 nm do you get a reasonable absorption. Otherwise you cannot say whether the substance is fluorescent or not.

The approach described in this book develops analytical procedures starting with detection – in this case UV detection. As a rule, of thumb you can say that with a value of $E_{1cm}^{1\%}$ >200 (ε approx. 5000 according to molecular weight) you can still detect approximately 1 ng of substance as a peak in the chromatogram. This also applies to Ibuprofen at a wavelength of below 230 nm (the value referred to in the table is at approximately 260 nm and not at the maximum absorption of <240 nm – not shown in the UV spectrum).

A hint: Endogenous substances in plasma contain various functional groups and conjugated double bonds. Therefore, practice has shown: the shorter the wavelength in the UV, the more substances in plasma are to be expected, which may chromatographically interfere.

Back to Ibuprofen: If our question is to determine about 10 µg/ml plasma and about 1 ng as a peak can be seen in the chromatogram at 220–230 nm, then we would only have to inject 0.1 µL plasma unto the column. Is it so easy?

Let us now turn to HPLC separation. The obvious approach is the RP separation (reversed phase, C 8 or C 18). In order to elute Ibuprofen you need methanol or acetonitrile (ACN) in higher concentration in the mobile phase. Thus, you are not in a position to inject plasma directly because the plasma proteins due to the higher share of organic solvent in the mobile phase would be precipitated at once in the column. Therefore, not only plasma dilution (for instance 0.1 µL plasma with water/1:20) but also sample preparation are necessary. The substances contained in plasma constitute a further problem: At 220–230 nm a larger part of them absorb comparatively well. Plasma itself contains approx. 8 % (= 80 000 ppm) plasma proteins – the larger share of it as Albumin. This shows clearly that at first many unwanted substances have to be eliminated before you can succeed in determination. A successful HPLC separation plays only a subordinate part in this case in the whole determination. Ibuprofen in its undissociated form tends to be rather lipophilic. This fact leads to the recommendation of an LLE (liquid-liquid extraction) with an organic solvent not mixable with water: Medium polarity like diisopropylether, chloroform or mixtures of hexane/heptane with middle polar solvents (see Section 3.3.1.2 for more information). For example, ethyl acetate would be easy to handle, however, it absorbs several per cent of watery phase, and therefore it can usually only be used in a mixture of aliphates in order to reduce the absorption of water significantly.

A rather effective precleaning can thus be reached as follows: After addition of, e. g., H_3PO_4 to the plasma and mixing thoroughly with an organic solvent (not mixable with water), you centrifuge the sample vial and take the (usually) upper organic phase subsequently, which then contains the majority of the Ibuprofen present in the sample.

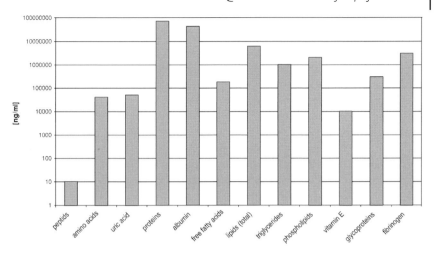

The concentration of selective substances in the plasma (attention: concentration to intervals of 10)

With this extraction not only can all hydrophilic substances be eliminated (thus also nearly all proteins, peptides and amino acids), but also all lipophilic basic substances because those are salts under these pH conditions and remain in the water phase. Thus, with the Ibuprofen all lipophilic acids and all lipophilic neutral substances are extracted. When this precleaning is not sufficient – the chromatogram shows too many disturbing substances at the elution time of Ibuprofen – then you can carry out a second extraction step: a re-extraction. In doing so the organic solvent is not evaporated but mixed with a watery basic buffer, shaken and centrifuged. Now all acids– including Ibuprofen – go into the watery phase as salts and can be – under certain circumstances – directly injected into the HPLC system. With this 2nd step of extraction all neutral lipophilic substances can be removed, they remain in the organic solvent. After such a precleaning you can very easily and even selectively gain a determination of Ibuprofen with HPLC-UV at 220–230 nm down to 10 µg/ml plasma. With fluorescence detectors, which have still enough light intensity at 220 nm, you can even get a far more selective detection for Ibuprofen[2]. An alternative sample preparation would be a solid-phase extraction (SPE): plasma slightly made acidic, is applied (preferably offline) to C 8- or C 18-material (solid-phase extraction cartridges) to which Ibuprofen adheres well. Next step: the column is eluted with methanol or acetonitrile after a washing step with water. Though the elute not only contains Ibuprofen, but of course also all lipophilic acids, all lipophilic neutral substances and also partly lipophilic bases.

A further important aspect with respect to sample preparation and clean up is the protein binding, which you must always pay attention to: most of the drugs and metabolites are subject to protein binding, a more or less strong one, mostly in the

2   Eur. J. Clin. Pharmacol. 48 (1995), 505–511 "Comparison of the bioavailability of dexibuprofen administered alone or as part of racemic ibuprofen"

manner of a Nernstsche distribution (a lipophilic gradient so to speak). As a rule, this protein binding can be dissolved easily and very quickly with liquid-liquid or with liquid-solid extraction (occasionally not so well). Ibuprofen's protein binding comprises more than 99 %. However, this is not important with the above-suggested sample preparation because the protein binding dissolves easily (see also Useful tip 6: protein binding).

Remark to determination in plasma and protein binding: An analyst can only give a determination of the total quantity of substance within plasma/serum in nearly all cases. Because the protein binding is a factor that varies and already slight modifications of plasma (dilution, change of pH, supplement of solvents) can change this factor dramatically, we must search for conditions that guarantee a safe determination of free substance together with protein-bound substance.

For instance, if you choose to eliminate proteins by precipitation with $HClO_4$ or TCA, then a big quantity of Ibuprofen (due to protein binding or being enclosed) would be precipitated, too. Thus, this determination would lead you to totally wrong results (far below the actual value). If you use protein precipitation with acetonitrile or methanol on the other hand, the protein binding is dissolved (as a rule of thumb), before the proteins are precipitated. Thus, such a protein precipitation, as far as its accuracy of recovery is concerned, is in most cases the right and safe choice. Concerning HPLC separation there is not much you must consider; of course the mobile phase has to be acidic in order to chromatograph Ibuprofen as an undissociated molecule. As a rule, you should achieve a k' value of 2–5. Please avoid the mistake that you think of retention time instead of k' values: each type of column has different dead times depending on the length of the column, column diameter and flow of mobile phase. The next decision to make is, if chromatography should be carried out isocratically or by using a gradient. This decision depends on numerous circumstances and will be discussed at length later (e. g., Section 4.1.1).

Useful tip 1: Plasma or serum? When we have a free choice, then we take either serum, which we let stand a period long enough before the centrifugation (the reason for this is to ensure that coagulation is largely completed) or plasma, that was centrifuged thoroughly. Otherwise, there will always be a risk of "fibrin clots" as we used to call them or "fibrin threads". These tend to clog up pipettes when pipetting and you lose sample volume when you transfer into other containers. Of course, you can do a second centrifugation of serum or plasma samples, but this means a further step in the procedure and is only useful if you can centrifuge in a rather strong way. The fact that you ought to wait for some time before you centrifuge full-blood (samples), may make you decide not to use human serum: On the one hand, it may distract the daily routine in hospital, on the other hand there are some substances whose successful determination demands that they are centrifuged quickly (some of them at 4 °C) and are frozen quickly in order to avoid decomposition. A general rule, which must be obeyed, is: the analyst is obliged to establish and decide after pretests how plasma/serum must be gained (time, anticoagulant, stabilizers to be added) and at which temperature it should be stored.

1.2
Second Question: Determination of Tryptophan in Urine

Certain therapies or conditions of illnesses change the secretion of certain hydrophilic acids into urine. Suppose, your task is to work on a method for detection of Tryptophan in urine.

According to our effective way of doing things we start again with questions as follows

- Which limit of determination is desired?
- Which kind of device is at your disposal?

For our given example we suppose that a limit of determination of 1 µg/ml urine is required and that, although an HPLC-MS-device (single quadrupole) is available in your laboratory, it is, however, needed for other research most of the time and thus cannot be used. Should this subject (determination of Tryptophan in urine) turn out to be effective, it might lead you to think of using that device, too. But for the time being you have got to do things merely with an HPLC-UV (variable wavelength/DAD) as well as an old electrochemical detector.

As a next step we consider molecular structure, solubility, UV spectrum and electrochemical behavior as well as MS-ionization (for future steps) of Tryptophan respectively.

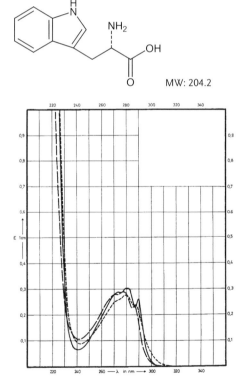

MW: 204.2

Table 1.2: Tryptophan – Details to the UV-Spectrum

solvent symbol	methanol ———	water – · – · –	0.1M HCl – – – –	0.1M NaOH · · · · · ·
absorption maximum	290 nm 280 nm		286 nm 278 nm	288 nm 280 nm
$E_{1cm}^{1\%}$	259 303		234 290	238 275
ε	5290 6190		4780 5920	4860 5620

- Structure: Tryptophan is definitely an amphoteric molecule – as are all amino acids – thus it is hardly extractable. It has got some lipophilic structure parts – so it is extractable with SPE (C 8 or C 18). It contains two essential functional groups: -NH 2 and -COOH (-NH in the ring tends to be subordinate).
- Solubility: generally speaking it dissolves well in water, alcohols and acetonitrile.
- UV spectrum: absorbing well at 280 nm (and thus it does not present a problem using gradient elution with acetonitrile or methanol).
- Electrochemistry: the possibility of oxidation of the -NH in the ring (technical literature contains reliable details on oxidation of Tryptophan).
- MS: probably good ionization yield under positive conditions (if necessary ask a colleague who has more experience within MS).

I cannot argue for mass-spectrometric detection – besides the fact that you have not got an instrument at your disposal – because ionization and detection with mass spectrometry yield as a rule less reproducibility and accuracy compared to UV detection unless using a chemically extremely similar internal standard to the substance (stable isotope labelled substance preferably). However, such an internal standard can be rather expensive and in cases of novel substances there might not be such a substance commercially available (for further information look up Section 2.3.6). Arguments against electrochemical detection: only a few analysts have still got experience with such a detector today. The handling of the instrument is distinctly more complicated than that of a UV detector, however, its selectivity and sensitivity are sometimes excellent. Besides, compared to MS its costs are very reasonable.

So we will try to solve the question with HPLC-UV. At 280 nm Tryptophan shows an $E_{1cm}^{1\%}$ of 303, i. e. 1 ng of substance can be easily detected as a peak and this at a distinctly more selective wavelength than for Ibuprofen determination in the last chapter. So we would have to apply 1 µL of urine or the substance out of 1 µL urine onto the HPLC column. Which column is recommended? As Tryptophan has a certain lipophilic structure, you can use C 8 or C 18 columns with 5–15 % methanol or acetonitrile. For safety reasons we recommend you to take an AQ-column (various manufacturers offer reversed-phase materials, which you can run in a watery phase for a longer period of time without the separation process collapsing being caused by lack of organic solvent). You could also use an ion-exchange column, whereby cationic exchangers (binding the -NH$_2$ of Tryptophan) are to be preferred to anionic exchangers (binding the -COOH of Tryptophan), as we have found out. However, because the use of and the experience with ion exchangers are rare in analytical laboratories, I rather recommend to you use C 8 or C 18 columns.

Sample preparation is hardly feasible (only strong RP materials like ENV+ (an offline SPE cartridge) or ion exchangers are suitable for it), thus it is necessary to inject urine directly. As a general rule with urine (especially deep-frozen urine), you have to be careful that all compounds are solved again after thawing (often it only helps to warm it to 37 °C and to shake it or to use ultrasound).

You must always consider to take a representative sample (which is distinctly easier if you have got plasma!). In order to avoid substances precipitating again you need to dilute 50 µL urine with, e. g., 250 µL water (perhaps you take 5 % of methanol in water, so that also more lipophilic substances remain dissolved). Then you could inject 10–20 µL unto the HPLC column.

As practice shows, two problems can arise:

1. If you succeed in gaining a selective method isocratically (that means a narrow peak appears). This peak only gains height if you add similar quantities of Tryptophan as a pure substance to your urine sample. But presumably in the next or at the latest in the 5th chromatogram of urine injection a problem will show up: broad signals of lately eluting peaks (from former injections) will disturb you in your determination. What you can do to counter this, is to apply a "gradient system" and after the isocratic separation to switch to an organic share of 80–90 % for 1–3 minutes, so that disturbing substances get rinsed out quickly from the column.
2. If separation is unselective (the peak of a pure Tryptophan solution in the chromatogram is narrower than in urine after adding a pure Tryptophan solution to the urine sample), you ought to repeat separation with RP columns of various manufacturers, in order to get a better selectivity.

An alternative and often better strategy for molecules like Tryptophan is to carry out an ion-pairing chromatography. I assume that you have already had to work with buffer systems when you used C 8/C 18 materials or TFA in the mobile phase, to obtain a good peak shape. In the given example you can use, e. g., pentane sulfonic acid (approx. 10 mM) in 10–20 % acetonitrile or methanol in the mobile phase (you need certain parts of organic solvents for solving pentane sulfonic acid and for the elution).

Pentane sulfonic acid and Tryptophan together form an ion pair under acidic conditions (using the -NH$_2$ of Tryptophan). This ion pair requires a distinctly higher percentage of organic solvent for elution than Tryptophan alone.

Are the disturbing substances only composed of organic acids derived from urine (which is probably the case), then these do elute much sooner as the ion pair and will not disturb the determination of Tryptophan any longer. However, I assume that an analogous step of rinsing out disturbing substances, which elute later and would disturb the determination of Tryptophan in a later chromatogram, by a gradient is necessary.

1.3
Third Question: Determination of Paclitaxel in Tissue

It is possible that you are requested to determine certain substances in different tissues. We presume that the analyst has gained sufficient experience with plasma and urine as matrix, but none with tissue. The decisive factors for determination in tissue are:

- homogenizing of the tissue;
- reaching a high recovery rate of the substance to be detected.

Useful tip 2: Tissue sample preparation   As we have already analyzed very many different types of tissues successfully, I would like to mention a few general rules here (for more details see Section 11.19).

You have to distinguish between tissue that is soft, nearly without fibers and other kinds of tissues. Soft tissues such as brain or often also liver, can be very well homogenized with an Ultraturrax. In doing so the procedure is as follows: After you have cut the defrosted tissue (it is not always easy to take a representative sample), a part of it is weighed into, e. g., a plastic centrifuge vial and mixed with 2–3 parts of ethanol, watery ethanol or acetone (depending on the substance to be determined and the matrix) and then homogenized with the Ultraturrax within 30–60 s (longer times can warm up the sample to higher temperatures). A homogeneous share of this blended tissue or this suspension is then weighed into, e. g., a small vial and diluted several times with water or certain hydrophilic solvents, so that you obtain approximately a 10 % share of protein – which is similar to plasma then. Then, use a procedure following the principles used with plasma.

For all other kinds of tissue I recommend to use, e. g., a so-called Dismembrator, which works in principle in this way: thus with an iron ball or a similar tool you pulverize the – due to liquid nitrogen – frozen tissue within about 30 s. There is rarely a tissue that cannot be pulverized with that procedure at such low temperatures.

The recovery of a given substance from tissue is always tested in two steps except for endogenous substances:

1. First, you add to a part of the homogenized blank tissue the substance in question in sufficient concentration (the concentration must not be near the determination limit, because otherwise the results would vary too much and would not be representative). Then apply your sample preparation. Do that also with a solution without tissue and compare the results. Only when you got a high recovery rate (to obtain that sometimes requires many and various pretests), can you take the next step.

2. If there already exists a tissue from preclinical or clinical application, you can vary the extraction conditions, which you have determined before with several similar homogenates to obtain a maximum yield.

A marginal note: in case these extractions ought to be analyzed with HPLC-MS or MS/MS you have to be aware that immense matrix effects can occur, which may suppress ionization of certain substances up to a remarkable 95 %. A brief discussion concerning an important matter: Irrespective of transferability of sample preparation from plasma to tissue the recovery from tissue may appear to be completely different (that means it may be worse). Occasionally, things turn out to be alright, when we used various detergents, if the recovery was a difficult matter.

There are nonionic and ionic tensides, relatively hydrophilic as much as hydrophobic ones. To apply tensides can bring forth a miracle, but pay attention to the fact that you get rid of the larger share of tensides, before you carry through the HPLC analytics. Large quantities of tensides can not only influence the HPLC separation, but also suppress the MS signal so much that you will not be able to recognize your substance of interest. In such cases especially it is crucial to determine, if recovery is still low or if strong matrix effects occur.

To sum up: Clinical analytics aims at obtaining safe and accurate results by applying methods that can be developed by good analysts in a relatively short time. This book shall be one tool to foster that ability.

MW : 853.9

Paclitaxel is a molecule with a medium polarity (for further details: requested determination limit, structure and solubility, see Section 11.14 or the Appendix). The tissue is cut into small pieces (take representative sample!), deep frozen for 15 min in liquid nitrogen with an iron ball contained in a small Teflon box and then smashed with a Dismembrator for about 30 s. A small portion of the sample (5 mg out of 30 mg, approximately) is mixed with 100 µL of 50 % ethanol, 20 µL of Triton X 100 (5 % in water) and 300 µL water; and then you can extract (LLE) with 4 ml diisopropylether. If you do not add ethanol you will have difficulty concerning the solubility of the substance to be analyzed and – without Triton X 100 (detergent) – the recovery with certain tissues will turn out poorly.

2
Planning of Analyses

2.1
Introduction

As already mentioned in the Preface there are some decisive questions that you
have to consider before you develop an HPLC method:

Question 1:   What limit of determination is needed (in plasma, urine or
tissue)?

Question 2:   What detectors are at your disposal?

Question 3:   What structure does the substance to be analyzed have?

Question 4:   What solubility does the analyte own (nondissociated and as a
salt, respectively)?

Question 5:   a) With UV detection: What $E_{1cm}^{1\%}$ value (ε or value) does the
molecule have at what wavelength?

b) Does the molecule perhaps fluoresce, which can be made
use of for determination?

c) Can the molecule perhaps be electrochemically oxidized?
(A reduction requires more knowledge and thus is difficult
to carry through.)

d) How sensitive could the MS or MS/MS detection of the
molecule be?

When you have answered these questions roughly, then you can undertake the
planning of the analyses.

Which steps do you have to consider?

Step 1:   What absolute detection limit (peak with a signal/noise ratio
of > 3:1) can be reached with the detectors that are at your
disposal?

Step 2:   What determination limit must be reached in the requested
matrix? How much of the matrix, as an extract or as a matrix itself (e. g., urine) has to be injected unto the HPLC column? After step 2 you know whether the detector chosen is
in a position to reach this determination limit in this matrix.
(These considerations do not yet include eventual disturbing

HPLC Methods for Clinical Pharmaceutical Analysis, First Edition. Hermann Mascher
© 2012 Wiley-VCH Verlag GmbH & Co. KGaA. Published 2012 by Wiley-VCH Verlag GmbH & Co. KGaA.

substances that could perhaps hamper the determination.) If it then appears that you need to inject, e. g., more than $100\,\mu L$ of plasma, this would imply that you need rather sophisticated precleanings and enrichment steps to reach the aim of a sufficiently sensitive determination. If your answers to the questions contained in step 2 are satisfactory, then you can start with step 3 – the practical sample cleanup.

Step 3:   Which solubility as an acid, base, salt or neutral substance does the molecule have?
Lipophilic molecules are more likely to be cleaned from the matrix efficiently than hydrophilic ones. If your calculation shows that you need less than $5\,\mu L$ of biological matrix for injection unto the HPLC column, then you need no enrichment step but only cleaning.

Step 4:   Consists of HPLC analytics, often carried out with RP (reversed phase) chromatography (for numerous details see Section 4.4.1 and other appropriate ones).

2.2
Limit of Detection (LOD) and Determination (LLOQ)

a.) For nearly all the drugs on the market you find information in the literature on Cmax values that can be reached in plasma after different application routes (orally, im, iv, sc) and with which elimination half-life these substances can be eliminated either due to excretion (urine, faeces) or due to metabolism (the Appendix gives approximately 100 examples). According to the guidelines of the FDA (Food and Drug Administration) for bioavailability or bioequivalence studies the limit of determination (LLOQ) should be at least at about 5 % of the expected Cmax value[3]. Drugs, however, show frequently varying C_{max} values due to patient/volunteer variations, different kinds of metabolism or variations of oral taking (before or after the meal), thus you should get a determination limit of about 1 % of the medium expected C_{max} value.

b.) Details on concentration of drugs as to be expected in urine are scarcely or not at all to be found in literature. If the drug in question is on the market already (and not in preclinical or in clinical – phase 1, 2 or 3 – development), you sometimes have the opportunity to obtain a couple of urine samples from patients (after application) and then you can – for orientation's sake – analyze these samples in order to define the concentration in urine roughly.

c.) Frequently, there is no information with respect to plasma or urine levels of phytopharmaka, or the information is not specific to substances, but to extracts only.

---

3   Guidance for Industry Bioanalytical Method Validation, FDA, May 2001

d.) With substances of the organism (endogenous substances) – if they are not really "exotic" – most details are available about healthy and ill persons[4]. The limit of determination due to practical reasons should be just under the lowest given value. If you find diverging information on concentration values of the substances in question, then you should examine the methods applied thoroughly and then separate "the wheat from the chaff".

2.3
Detectors

2.3.1
UV Detectors

No matter which type of UV detector you use, whether the old-fashioned fixed-wavelength detector at 254 nm or a detector of variable wavelength at, e. g., 190–600 nm or a DAD (diode-array detector), you cannot determine less than 0.1–0.5 ng of a substance as an absolute amount (per peak) (and even that with a suitable UV spectrum only, i. e. if it is not necessary to choose a very short wavelength). Generally speaking, you rarely will be able to detect less than 1 ng absolute amount of a substance. When you apply UPLC devices and HPLC columns of 2 mm inner diameter you can take as a guideline for the peak of the substance approximately 0.1 ng. To use UV detectors is especially recommended with substances that have UV maxima at wavelengths higher than 300 nm because neither the components of the mobile phase matter (this applies to common solvents as well as to common buffers) nor do many disturbing substances in the matrix (no matter whether you have got plasma, urine or tissue) absorb at such wavelengths any longer.

According to our experience with sample enrichment from plasma or urine only about 5–10 ng/ml can be reached as determination limits exist as a general rule.

If you are unlucky it can be much higher, even with a molecule with a good $E_{1cm}^{1\%}$ value (ε value) at a higher wavelength (>280 nm) and also being rather lipophilic (therefore extractable). Nevertheless, in such cases you can only get an LLOQ of, e. g., 50 ng/ml, if some endogenous substances with similar behavior as the analyte are in the sample.

2.3.2
Fluorescence Detectors

This kind of detector is an inexpensive alternative for sensitive and selective determinations, if you do not posses a MS or MS/MS detector. As is known, there are only a few substances that possess useful fluorescence. This is useful inasmuch as your analyte shows fluorescence, anyway, and the excitation and emission wavelength are far enough apart so that you do not have strong stray light of the excita-

---

4   See list of books: Geigy tables

tion wavelength (through Rayleigh, Tyndall and Raman effects). Through our intensive practice there were sometimes molecules that had almost no or even weak fluorescence, but under extreme conditions (e. g., pH values) the fluorescence was quite good. Occasionally, not only the pH of the mobile phase was important, but also the kind of buffers used and the percentage of organic modifier.

Besides, fluorescence is an especially good tool to be used after derivatization because of high sensitivity and often reasonable selectivity. This means that specific functional groups in the molecule can be transformed into products with good fluorescence relatively easily by using precolumn (offline) or postcolumn (online) derivatization. Chapter 6 deals with this topic extensively.

2.3.3
Electrochemical Detectors = ECDs (Amperometric, Coulometric)

Two types of ECDs exist: amperometric and coulometric. Amperometric detectors often use glassy carbon electrodes that have an efficiency of 1–5 %. This means 1–5 % of your analyte are oxidized during passing the detector in the mobile phase. Quite different The coulometric detector is quite different – working with porous graphite that is run through by the mobile phase.

Your analytes can be oxidized almost quantitatively under certain circumstances, so a second electrode can be used for another oxidation step just behind the first electrode. This enables you to take a higher potential for other substances or you reduce with the second electrode the substances oxidized at the first electrode. Both types – amperometric and coulometric – are great tools for sensitively oxidizing different substances. Well known is the most sensitive detection of catecholamines from plasma down to low pg/ml plasma levels. Generally speaking, all phenols can be determined, but also other substances that are easily oxidized such as ascorbic acid, purines, vitamin A or E and also different aromatic amines. A drawback of using an ECD is that you have to clean such a detector often and you have to be careful during the preparation of your mobile phases. Furthermore, gradient elution is almost impossible for sensitive determinations – a bad baseline shift occurs very often.

There are also specific pulsed ECDs on the market usable also for, e. g., sugar determination with gold cells, switching from positive to negative current within milliseconds. But all the ECDs do not belong to the standard equipment of bioanalytical laboratories.

2.3.4
Refraction Index Detectors

These detectors have hardly any field of activity in clinical analytics because their absolute detection limits are relatively high. Besides, their selectivity is extraordinarily low.

2.3.5
Mass Detectors (ELSD, etc.)

These detectors are used extremely seldom and very specifically for certain classes of substances as cholesterol derivatives or sphingolipids. They are used for substances with a strong tendency to be nonvolatile – this is even a prerequisite – with no usable chromophores. Their sensitivity and selectivity is relatively low.

2.3.6
Mass Spectrometers (MS or MS/MS)

MS (single-quadrupole-instrument): the substance positively or negatively charged is selected in the quadrupole after the evaporation step and detected. These SIM measurements can definitely be sensitive (low to medium picogram range in absolute terms), but for measurements from matrix (e. g., plasma or urine) in the lower determination range their determination is particularly unselective. Therefore, higher LLOQs result due to lack of selectivity.

MS/MS instruments are more sensitive and much more selective. They are divided into two types:

- The so-called triple-quad instrument can be used very well for quantification in MRM (multiple reaction monitoring). The detection limits (LOD) for substances easily ionized are at best at 10–20 femtogram per peak for highly sensitive instruments.
- The ion trap (another type of MS/MS) is very reliable to get information on molecular structures, but it is less useful for quantifications (especially with larger calibration ranges).

But both types of instruments (MS/MS) also show the drawbacks of extreme selectivity: Strong matrix effects can easily occur without any visual hint in the highly selective chromatogram (often ion suppression, sometimes enhancement of the signal). When using MRM disturbing substances can only be recognized with difficulty, each disturbing substance can easily be shown by the UV detector in the chromatogram. Using UV you can hopefully evade the problem while developing the method.

2.4
Structure of the Analyte

It is necessary to have a close look at the structural formula of the analyte – from all perspectives. A few examples shall show clearly what is critically important.

2.4.1
Functional Groups

First, acids (mainly -COOH) and bases (-NH 2, =NH, ≡N) must be mentioned because these groups are of special interest, both for preparing the samples and the
chromatography. Phenols (aromatic -OH) may play apart also at pH >9.5 because
then they exist as hydrophilic phenolates. Pay attention to thiols (-SH), because
they are easily oxidized often creating -S–S- bonds and thus are withdrawn from direct determination. Alcoholic groups (primary, secondary and tertiary -OH) influence the hydrophilic/lipophilic properties of a molecule and may be important for
derivatization. Ketones and aldehydes are rather insignificant for sample cleanup
and chromatography. Sometimes, they can be useful for specific derivatization.

The derivatization of primary and secondary amines (pre- and postcolumn) is often a very good possibility (e. g., for amino acid derivatization). The carboxy group
is less suitable for derivatizing, especially for HPLC derivatization, than amines.
This is because useful derivatizing means especially esterization (e. g., methylation) exist for GC and not HPLC. Such reactions are good for enhancing volatility
but not for UV or fluorescence detection used in HPLC. Thiols can partly be derivatized in a very useful way to protect them from possible breakdown and also to
enhance detection sensitivity[5].

2.4.2
Lipophilic Basic Framework

A molecule cannot only have functional groups, but also a more or less lipophilic
basic framework. Whether this basic framework plays a crucial part depends very
much on the functional groups.

Unpleasant for any type of selective sample preparation are amphoteric molecules or molecules similar to sugar or containing sugars (e. g., flavonolglykosides).

2.4.3
Influence of the Structure on the Behavior of a Molecule

At times it can be very helpful to have a close look at the special arrangement of
a molecule, if certain behavior during the extraction or chromatography process
does not comply with the usual rules.

Let me present and discuss a few examples:

5   See list of books: Lingeman

- Ibuprofen

The most striking features of this molecule are the -COOH group and the large lipophilic remainder. The lipophilic properties make the extraction into organic solvents (LLE) or the retention on SPE columns (solid-phase extraction) possible. But the prerequisite is always an undissociated carboxyl group. If the pH value is neutral or basic, this group and not the lipophilic remainder define the behavior of the molecule.

- Phentermine

The counterpart of Ibuprofen is Phentermine: a basic molecule (-NH$_2$) with a large lipophilic remainder. The performance is analogous to that of Ibuprofen. The lipophilic properties become noticeable only if the basic function is not dissociated. At a neutral pH or acidic pH value, however, the basic group provides water solubility and the lipophilic remainder has low influence.

- Phenylsalicylate

This molecule is dominated by the phenolic group (aromatic -OH). This group is insignificant at pH values smaller than 9, but dominates the molecule as phenolate at values greater than 10. In a basic environment the water solubility thus dominates.

- Captopril

Generally, it is a very hydrophilic molecule, its behavior is dominated by the COOH group. In biological matrices, though, the extremely high oxidability of the free SH-group (thiol) plays the most important part. Especially -S–S bonds withdraw a molecule of this type very quickly from determination. With dithiothreitol (DTT) or tributylphosphine for instance molecules with free SH groups can emerge and thus can be detected again.

- Phenylalanine

Generally speaking, it is a relatively hydrophilic molecule (amino acid) where the lipophilic character only occurs, if either the acid or the basic group is bound, e. g., to other amino acids building a peptide. With amphoteric molecules like Phenylalanine there is little freedom for usual sample preparation procedures. A good way could be cationic exchangers (binding of -NH$_2$) or anionic exchangers (binding of -COOH).

- Rutine

A typical molecule of a glycoside coming from plant substances is Rutine, possessing some phenolic groups (behavior like Phenylsalicylate). The solubility and the chromatographic behavior are very strongly influenced by the two sugar moieties. Thus, in the end Rutine is a hydrophilic molecule.

The following habit has proved its worth: To really get into the structural formula of a molecule with regard to sample preparation of the specific molecule (what is beneficial, what can be "dangerous"), but also with regard to chromatography. But detection – as mentioned – is always the first part to consider!

2.5
Solubility of an Analyte

If you have considered the structure of an analyte (see Section 2.4) intensely, you should then know a lot about solubility. Some details – especially for LLE – you get often first from the literature. It is important to distinguish the solubility of a salt on the one hand and of the undissociated molecule (much more important) on the other hand. Sometimes, such details are helpful for preparing stock solutions with relatively high drug concentrations because often concentrations of 1 mg/ml or more are necessary for stock solutions. Ensure that all of the substance is dissolved, even after storing at –20 °C and thawing. Forming salts is often helpful to get such high concentrations solved. Please do also ensure with HPLC injection solutions that everything (of your analyte and internal standard) is solved. The percentage of organic solvent plays an important role in injection solutions: Everything must be solved (often higher organic parts are necessary) as well as the elution power of the injection solution should be lower than the elution power of the mobile phase especially by means of injecting high volumes.

2.6
Selection of the Detector

In Section 2.3 some aspects of different detectors are shown. But at this point you have to choose which detector you want or you can use: If your determination is not meant for submission to the EMA (European Medicines Agency) or the FDA (Food and Drug Administration), we suggest using the simplest detector suitable for the topic at hand. It should be the cheapest variant and also the simplest variant for operation, if it fits the determination purpose. HPLC-UV – if the sensitivity and selectivity matches your topic – can give you the best precision and accuracy, better than MS or MS/MS. It is the same with fluorescence detection compared to MS or MS/MS. Besides, you can go without an internal standard more easily by using UV or fluorescence detection than by using MS or MS/MS. Sometimes, it is not easy to find an appropriate internal standard. But compensation of varying recovery rates and of complicated sample preparation often needs an internal standard anyway, no matter what the detector is.

Final thoughts to complete the picture and sum up:

How deep could the detection limit (LOD) with one's own detectors be (also see Section 2.3)?

- *UV detector:* Decisive for the detection limit are two factors: the $E_{1cm}^{1\%}$ value (or ε value) and the wavelength of the UV maximum. With an $E_{1cm}^{1\%}$ value of >700–800 you can detect even 0.2 ng absolute as a peak, with <250 nm many endogenous substances can be disturbing. Lower than 220 nm even the mobile phase through baseline drift or a noisier baseline can disturb the determination. If the maximum is >300 nm, your molecule is very suitable for UV detection, if the absolute detection amount is enough. Sometimes, it could be helpful not to use the maximum but a higher wavelength with sufficient absorption (this is always good for selectivity reasons!)
- *Fluorescence detector:* If you are lucky to determine a fluorescent molecule, then often everything can go well. Prepare some real samples (often protein

precipitation is enough with a strongly fluorescing analyte), check the recovery, optimize before starting with some fast runs the optimum excitation and emission wavelength (not always at maximum because of 2 reasons: signal-to-noise ratios are important – observe stray light – and sometimes because of selectivity reasons due to endogenous substances). Usually, the highest wavelength usable for excitation is the best wavelength because of selectivity. Keep in mind all the stray light coming from the excitation like Raleigh, Tyndall and Raman effects.

- *MS and MS/MS:* It depends very much on the molecular structure which ionization you could or would use. Usually, someone has ESI (= electrospray ionization) and APCI (= atmospheric-pressure chemical ionization) at hand – if at all an MS. Sometimes, APPI (= atmospheric pressure photo ionization) could be also a good option for highly lipophilic molecules or sulfur-containing molecules.

3
Sample Preparation

Generally speaking, there are many different ways for sample preparation in the bioanalytical field – none like the other. Nevertheless, some general concepts exist that can be adapted in a flexible manner to each topic (matrix, substance, separation, detection).

3.1
Dilution

The simplest way for sample preparation is dilution. There are indeed some topics where the concentration of the analyte is so high compared to the detector sensitivity that a single dilution step is enough. Usually, only urine samples can be diluted because of not having such high protein concentrations compared to plasma. Plasma proteins would be precipitated in the HPLC column by using a gradient in RP-chromatography or using isocratic conditions with more than 8–10% ACN or 15% methanol. In former times we indeed analyzed Tyrosine and Tryptophan in plasma only after dilution with water at around 10% methanol in the mobile phase with ECD detection.

3.2
Protein Precipitation, Overview

With plasma samples there is an option to precipitate the proteins with strong acids like $HClO_4$ or TCA (= trichloroacetic acid) or with ACN or methanol. In the literature there are other solutions (different salt solutions) or solvents (e.g., ketones) mentioned for protein precipitation, which have no common use.

Protein precipitation always means that all low molecular weight substances will stay solved (including all small endogenous substances and peptides up to roughly 10–15 kDa). Therefore, all these substances can disturb your determination. Having success means either high concentrations of the analyte or very selective and sensitive detection – often in combination with a good HPLC separation.

HPLC Methods for Clinical Pharmaceutical Analysis, First Edition. Hermann Mascher
© 2012 Wiley-VCH Verlag GmbH & Co. KGaA. Published 2012 by Wiley-VCH Verlag GmbH & Co. KGaA.

3.2.1
Protein Precipitation with Different Acids

Protein precipitation with strong acids (e.g., 1 ml plasma plus 0.2 ml 20% TCA or 20% $HClO_4$) is very efficient, but there are some drawbacks. The most important is that using this sample preparation only makes sense with substances that are not or barely protein bound. It can also be used with some restrictions with substances that have a low protein binding and almost no variation between different subjects. Furthermore, the substances must be stable under strong acidic conditions. Nevertheless, not all automatic sample injectors and all materials in HPLC columns tolerate such high acid concentrations. With substances with high protein binding the protein-bound part will be eliminated with the precipitated proteins. That a determination can still be successful is shown by an example from our lab from former times: By using an old HPLC-UV method for the determination of Doxycycline in plasma it was necessary – because of sensitivity – to use this kind of protein precipitation (because of peak shape and using a high injection volume due to a low detection limit). From the recovery we could calculate the total amount of the substance in plasma (about 40% recovery because of 60% protein binding).

If you have no or only little (not variable) protein binding you could use this kind of protein precipitation (provided that the drug is stable and you have an internal standard). After adding the acid you could vortex for some seconds and then centrifuge the sample. The clear supernatant definitely contains no protein. By using reversed phase columns the supernatant is so high in water content that you can inject large volumes without any peak broadening. Doing that only using a UV or fluorescence detector makes sense. An electrochemical detector is not an option because of too high an ion concentration in the injection solution that blinds the electrodes at least for a short time. Using MS detection is due to the large amount of acids – especially if it is $HClO_4$ – only possible if you split off the injection peak to waste and afterwards the eluent from the column is allowed to go into the mass spectrometer.

3.2.2
Protein Precipitation with Acetonitrile

Protein precipitation with acetonitrile is a common method for plasma sample preparation. By doing so usually all protein-bound analyte is set free. Acetonitrile is usually used at least in the ratio of 1:1 to plasma. But by using that ratio there would be a danger that not all proteins are precipitated. Sometimes, if you have not cooled down the precipitated sample for half an hour to 2–6 °C then the supernatant after centrifugation is still not clear.

Sometimes, a portion of the proteins will precipitate after a while, which might clog the injector or lead to an overpressure in the column. But why should you not use more acetonitrile (e.g., twice the volume)? Certainly, this could be a very good solution if the high percentage of acetonitrile is not a problem in the injection solution that could lead to broad peaks if larger volumes are injected. But you can reduce that high percentage of acetonitrile by using a vacuum centrifuge for gentle and well-controlled evaporation of acetonitrile. So, you can get a highly aqueous injection solution. This step makes it necessary – because of variable volumes – to use an internal standard.

Useful tip 3: Percentage of organic solvents in the injection solution

Some remarks on a high content of acetonitrile and also methanol in the injection solution by using reversed phase separation: As we know, analytes will be eluted faster by using a higher content of organic solvents in the mobile phase by using isocratic elution conditions when injecting the supernatant of the acetonitrile precipitation of plasma (2:1 which means approx. 67% acetonitrile). If you use 20% acetonitrile/80% water as a mobile phase only small volumes can be injected without peak broadening (otherwise a so-called "leading" occurs that means peaks go up slower than they drop). Naturally, this effect also depends on the column diameter and eluents. Altogether, it is important that 20–50 µL of 67% acetonitrile change the separation conditions at the column entrance in a way that the mobile phase has a stronger elution power for a short while (It is a gradient from 67% acetonitrile to 20% acetonitrile for a short time). Therefore, all substances at the column entrance are eluted faster for a short while and then curbed, thus resulting in a broader peak.

If you have the opportunity to evaporate acetonitrile in a vacuum centrifuge, this can solve your problem, although only by using an internal standard because of varying volumes. With acetonitrile a bit of water will be evaporated too in varying volume (depending on time and vacuum). This method is also not suitable for highly lipophilic substances. These substances are no longer solved in the resulting pure watery injection solution. At the latest though, if you freeze the injection solution before injection and inject the samples, e.g., a day later, a substantial part of the lipophilic analyte will precipitate and cannot be solved again even by using a vortexer or ultrasonic bath.

3.2.3
Protein Precipitation with Methanol/Ethanol

Protein precipitation with methanol leads to higher recovery rates for specific molecules – although only seldom. Usually, we suggest four times more methanol than plasma, resulting in a supernatant of about 80% methanol in the injection solution, which could be a problem because of high elution power (see Useful tip 3).

3.3
Extraction

There are usually two different ways of extraction: LLE (liquid-liquid extraction) and SPE (solid-phase extraction). Both ways are useful – depending on the analyte – to clean samples and/or perform sample enrichment.

3.3.1
Liquid-Liquid Extraction (LLE)

In former times LLE was an important step to clean up samples coming from synthesis. The principle: the analyte can be extracted due to its lipophilic behavior from water (plasma, urine) into an organic solvent. It is easy to add an organic solvent (not soluble in or mixable with water) to plasma and – after combining, e.g., 1 ml plasma in a centrifuge tube with 4–6 ml organic solvent – to mix it vigorously for 1–2 min and to centrifuge afterwards to get phase separation. Most of the organic solvent (usually the upper phase) will be drawn out by a pipette. A very elegant procedure is to freeze the whole sample after centrifugation at –70 °C to –80 °C for about 10 min. After that the organic phase (only possible when it is the upper phase) can be decanted almost quantitatively in another centrifuge tube. If you have no freezer with such a low temperature you can use a beaker with acetone and dry ice to freeze the sample within 1 min. Keep two things in mind: The beaker must be significantly lower than the centrifuge tube, otherwise CO_2 gas will evaporate and spill over into the centrifuge tube. This can change the pH of the solvent (going to more acidic) and occasionally this could be hazardous. Secondly, you must clean the centrifuge tube of acetone before pouring the solvent into another empty centrifuge tube, otherwise your sample will be contaminated with acetone and CO_2.

Usually, the collected organic solvent will be evaporated afterwards. But another way could be to take another cleaning step, e.g., of re-extraction after changing the pH of the water phase for acidic or basic substances. Some more words on evaporation: It is clear that only such analytes can be treated in that way if they do not evaporate (also not as undissociated molecules). Two types of evaporating instruments can be used. If you have volumes of more than 1 ml organic solvent you should use a so-called Turbovap. The temperature of the water bath should be chosen such that the solvent can be evaporated within 5–15 min (as low as possible, as high as necessary). Some analytes are temperature sensitive, therefore use low temperatures for evaporation in these cases. It is important that there is not too much solvent in the vial so that by using nitrogen or compressed air no solvent will be spouted (otherwise reduce the gas pressure!). On the one hand, there could be less sample volume that would mean a lower concentration (if you use no internal standard) and on the other hand and more importantly samples with higher concentrations will contaminate and distort samples with lower concentrations or samples with no analyte (e.g., predose samples).

By using volumes below 0.5 to 1 ml of organic solvent you may use a so-called Speedvac (vacuum centrifuge). Going in this direction means that you may use your HPLC injection vials so you can save one operation. You then only need to resolve the residue of the sample in the HPLC injection solution (remember to vortex sufficiently!).

Sometimes, it might be difficult to resolve the dried residue in a small volume of injection solution after evaporation. It is important to redissolve your analyte and the internal standard – it is not necessary to solve each lipid. Therefore, if a cloudy solution results, this is not a problem if there are not really unsolved particles in that suspension that can block the frits of the analytical column. You can observe that if you inject and see a small pressure drop after each injection.

Useful tip 4: Reuptake after evaporation
The reuptake could be a real challenge as you have to consider many factors:
   a.) Solubility of the analyte/internal standard.
   b.) Solubility of higher amounts of different lipids coextracted (often fat, phospholipids, etc.).
   c.) Using as little as possible of injection solution (but as much as necessary).
   d.) The percentage of organic solvent (acetonitrile or methanol for reversed phase) should not be too high. Otherwise you can only inject very little volumes without loss of separation power. If you use too little percentage of organic solvent there will still be a layer of the evaporated sample sticking at the bottom of the glass wall in the vial including analyte and internal standard.
Most of these problems can be solved by using 50–100 µL DMSO before evaporation of a sample (this is not always possible!). DMSO is a really good solvent for many substances – even higher lipophilic – and DMSO is not easy to evaporate. Therefore, DMSO stays in the vial although the other organic solvent is evaporated. It prevents the sample from total evaporation. After evaporation of the organic solvent usually water, water–methanol mixtures, methanol or acetonitrile is added to the sample, followed by 30 s of vortexing and then transfer to the injection vial.

Sometimes, a first LLE step can be used to remove disturbing substances without extraction of the analyte. With basic analytes the sample (plasma/urine) should be made acidic, resulting in the analyte being dissociated in the matrix. Then, you can extract all lipophilic acids and neutral substances and the organic solvent for extraction can be disposed. Afterwards, the watery matrix without many lipophilic substances will be made alkaline and then the analyte can be extracted much more selectively. For lipophilic acids the plasma/urine will be made alkaline and extracted. With this step, all lipophilic basic and all lipophilic neutral substances were extracted, the extraction solvent can be disposed. Afterwards, the watery matrix will be made acidic and the lipophilic acidic analyte will be extracted much more selectively.

In a third extraction step the resulting organic solvent that contains the analyte, can be cleaned further: If the analyte is a basic substance the organic solvent could be re-extracted by an acidic watery phase. By doing so, the analyte will go as a salt into the water phase and this phase will only be spoiled with lipophilic basic substances that are inside as salts. The resulting organic phase contains only lipophilic neutral substances. If you look for those substances you can use this organic phase. If the wanted analyte is an acidic substance you can extract the organic phase with a basic watery phase, whereas the acidic analyte will go as a salt into the water phase. Still, all neutral lipophilic substances will stay in the organic phase – if you look for them.

These described extraction steps may lead, as a relatively simple process, to rather clean extracts for lipophilic bases, acids and neutral compounds.

But be careful! Your analyte and the internal standard must be completely stable under these acidic and basic conditions.

Either aqueous basic or acidic extracts sometimes can be used for direct injection. The only demand is that the analyte and the internal standard are stable under these conditions for hours in the autosampler. A good chromatographic separation is only possible under such circumstances if the analyte – which exists as a dissociated ion in the injection solution – is converted within a second in the column to the undissociated molecule (this is valid for RP chromatography). The same process takes place when within a second an ion pair is formed with an opposite charged lipophilic ion in the mobile phase. With acids as analytes this rapid change takes place with strong acids in the mobile phase (e.g., phosphoric acid, trifluoroacetic acid, perchloric acid). Salts of basic analytes can be converted by using higher concentrations of ammoniumcarbamate (e.g., 50 mM; but not every stationary phase is suitable for such a high pH) or by using higher concentrations of a counterion, like sulfonic acids (e.g., pentane sulfonic acid, but do not forget using some per cent of organic, e.g., methanol, acetonitrile in order to solve this lipophilic counterion).

Which solvents are really well fitting for LLE? Toxic solvents like benzene or carbon tetrachloride are really useful and were used in the past, but are not in use any longer.

3.3.1.1
Criterion for Fitting Solvents (Volatility/Toxicity/Solubility in Water/Relative Density)

Useful tip 5: Criteria for a fitting extraction solvent
Criteria for a fitting extraction solvent could be defined as follows:
   a.) *Good volatility* because the solvents often will be evaporated after extraction (when using re-extraction of the analyte into aqueous phases this is not relevant).
   b.) *Low toxicity* because with each sample 1–10 ml of this solvent will be evaporated (even if it happens in the extracted hood it is bad for the staff and the environment).
   c.) *Low or no solubility in watery phases*. Critical points are middle polar solvents not miscible with water but that are able to absorb 0.5–5% water. Such solvents are good compromises between good solubility of analytes (nonpolar solvents are often bad solvents for analytes) and for the absorption of water. The absorption of water using middle polar organic solvents can be reduced to an acceptable level by mixing with nonpolar solvents. Sometimes, nonpolar solvents would be optimal concerning phase separation after centrifugation without water absorption. But many substances are not lipophilic enough to be solved well in such nonpolar solvents. By the way – solubility alone sometimes is not the decisive parameter. There are some lipophilic drugs or endogenous substances (e.g., Cholesterol or Caritinoides) that are so strongly protein bound in a specific manner that recovery from plasma is poor. In such cases you need detergents, specific alcohols (e.g., butanol, isoamylalcohol)

in low percentage or the so-called "Folch extraction" (see below, chloroform/methanol) to solve such problems.

d.) An important point is also the *specific weight* of an organic solvent not mixable with water. Many solvents are lighter than urine (= water, density around 1 g/ml) or plasma (density around 1.025) and form – after mixing and centrifugation – a clear upper phase. Dichloromethane or chloroform are excellent organic solvents for many substances but they will be the lower phase. So pipetting is almost impossible. The only chance is using a syringe. With such procedures the danger of incorrect sample preparation or carryover are high!

Which solvents and mixtures of different solvents have proven successful?

- *Ethyl acetate*: Much lighter than water, resulting in the upper phase. Very useful for relatively hydrophilic substances that could just be solved in organic solvents, but absorbs a lot of water (resulting in not very clean extracts) and will be partly destroyed under strong basic conditions of the water phase. The resulting acetic acid is buffering unintentionally heading towards neutral pH.
- *Isopropyl ether*: Much lighter than water therefore in the upper phase. Very useful for medium-polar substances. Ethyl ether that was often used in the past is well known to form peroxide and evaporates easily under warm conditions in the lab in summer time. Consequently, it is not pleasant to work with for lab workers.
- *Chloroform/methanol (2:1)*: This mixture is used for the so-called "Folch extraction", this means to solve lipophilic substances (e.g., sterols or even vitamin A) from specific protein bindings (or other types of binding) and extract them because of the lipophilic behavior of the molecules set free. The procedure consists of mixing one part of plasma with three parts of chloroform/methanol. One phase is resulting from this ratio of different solvents that solve lipophilic substances from those specific bindings. Then, you have to add water or watery buffer solutions to get a two-phase mixture. After centrifugation the lower phase with the analyte consisting of mainly chloroform will be used for further clean-up or evaporation.
- *Hexane/heptane*: Depending on the required volatility one can use either one or the other option, but commonly the pure aliphate is not used. You may use the aliphate as a mixture with isopropyl ether, ethyl acetate, with a few parts of iso-amyl alcohol or also as chloroform mixtures. By using chloroform one should use hexane instead of heptane. The resulting mixture should have a specific weight lower than 1 so that the organic phase remains the upper phase.

3.3.2
Liquid-Solid Extraction (SPE = Solid-Phase Extraction)

In the last two decades SPE has been a fast-growing technique in sample preparation. In our opinion SPE has surpassed LLE already.

Which advantages does SPE offer?

a.) It is automatable.

b.) In general, less sample manipulation is required (at least), most labs can
handle this technique better than LLE.

c.) SPE often works with molecules where LLE as an extraction technique is not
possible (e.g., amphoteric molecule) or not longer possible (e.g., flavonolglycosides) with good results. Such molecules have sufficient retention on RP
systems because of lipophilic parts in the molecules.

d.) By using ion exchangers especially basic substances from different matrices can be extracted. But despite RP clean-up some more thoughts are
necessary for using ion exchangers. An eventual saturation of the sulfonic
acids or quaternary amino groups by proteins. Also, high or varying salt
concentrations in different urine samples can partly elute the trapped
ionized analytes. In general, the pH of the used matrix has to be in the
appropriate range but without using high concentrations of ions to fix it.
Many cleaning procedures can be chosen concerning organic solvents. This
effect is totally different to clean-up procedures in RP chromatography or
RP precolumn cleaning. For RP material such conditions are used to elute
substances. To elute substances from ion exchangers again needs much
brainwork. First, consider the necessary pH and ion concentration of the
elution solvent and do not forget: The use of, e.g., high concentrations of hydrochloric acid has a great deal of elution power but it can destroy the separation material as an injection solution for HPLC. The same problem could
be high ion concentrations in the elute by using HPLC-MS (the system may
become blind or as an alternative you can direct the HPLC effluent to waste
for a time and then switch to the detector before eluting your molecule).

e.) You can use SPE either offline or online. With offline you have an unre-
stricted option to use different solvents, salts, acids, bases. Online the freedom for different clean-up steps is very restricted (except by using expensive
automatic column switching units).

For lipophilic molecules we often use LLE and not SPE because of much higher
selectivity of LLE.

As described in Section 3.3.1 after finally three extraction steps exclusively only
lipophilic bases, acids or neutral compounds are in the injection solution. Besides,
the composition of the injection solution can be varied freely (see Useful tip 4).

In our opinion one important reason for people working in this analytical field
not using LLE is that they know too little about this. They have no practical impression or they believe it does not work for routine analytics. We disagree from our
practical point of view: By using our racks for sample preparation 50 vials can be

handled together compared to SPE. After pipetting plasma samples individually, the use of semiautomatic pipettes for the internal standard solution (and if necessary) for buffers and organic solvents is possible. After capping the vials by hand the whole rack could be shaken for a short time on an automatic shaker or – after using a second rack for covering – 1 to 2 min strongly by hand.

A centrifuge often has space for 48 vials and could be filled easily by hand. After centrifugation it can easily be emptied. Also, it is a procedure that can be done without a great effort. Even by using two or three extraction steps and freezing-out procedures one laboratory worker can prepare 150–250 samples per day. Using SPE offline the number would be similar.

3.3.2.1
Reversed-Phase Phases for SPE

Reversed-phase materials are predominant for many applications either online or offline. An important advantage of these phases is that many highly protein-bound substances (e.g., NSAR like Ibuprofen with more than 99% protein binding) are completely bound to these RP phases. Most of the proteins will go through the precolumn while breaking the protein binding within seconds (like with LLE). When plasma proteins pass through these small columns of 20–100 mg RP material through slight centrifugation or vacuum, the free analyte (this means not protein bound) will stick within seconds on the RP surface. We often use light centrifugation for this step because 48 simple samples can be handled together. The free part of a drug as well as the protein-bound part will stick to the surface of the filling material in the SPE cartridge. The protein-bound part will be similarly dissolved form the protein within seconds.

Useful tip 6: Protein binding

In the past there were many discussions about protein binding in our surroundings. There are colleagues who think that the forming of protein binding, respectively, dissolving from protein, is a slow process. Often this was a point of discussion when using ultrafiltration or especially using dialysis. This attitude of different substances on reversed phase columns, described in the last chapter, points out precisely – in our opinion – that the common protein binding is an extremely fast process – binding and splitting off!

Of course there are rare cases where – simply expressed – another mechanism of protein binding can occur that has nothing to do with the Nernstsche partition rule (solubility; octanol–water partition). There are other rules working that can be seen also when lipophilic molecules only partly stick on RP material.

In the whole field of RP materials there are sold many different products. The whole development of HPLC separation material you can find also in the area of SPE. From old material with strong rest silanol effects – strong surface effects especially for basic molecules – to polymers with neutral behavior you can find everything. From rough materials with 40–60 µm down to 5–10 µm materials there is a broad variety. Small particle size has naturally a certain backpressure.

One point you should keep in mind when using RP materials: There is almost no chance to get a good selectivity. Therefore, you don't need a great variety of RP materials. There is almost no difference for the user when using material from one to another manufacturer for common applications. By using different RP materials you can't win more selectivity for similar substance classes in general.

Proteins, salts or highly hydrophilic substances can be eluted as known partly easily even with water. And therefore they can be eliminated. This effect is, at the same time, a cleaning step for lipophilic molecules. And if it is not necessary to use pure methanol or acetonitrile for elution – when using offline clean up (many highly lipophilic substances will stay on the surface and are therefore eliminated) – a cleaner extract will result. All the eliminated highly lipophilic substances now have no chance to elute from the analytical column as late eluting peaks (sometimes in the following chromatograms). Usually, the clean-up step and the analytical separation will take place on a RP material therefore no or little selectivity enhancement will result by using RP material for sample clean up. Nevertheless, a huge preconcentration step is possible with such a combination, which really can be helpful for low concentrations of analytes determined in plasma. As nowadays the separation power of 2–3 µm materials in HPLC columns is very good, separation of a lot of possibly endogenous interfering substances from analytes and internal standards can be achieved. This is true especially for the determination in the ppm and ppb range. By determining even lower concentrations you will get many problems by using UV detection through many unwanted detected peaks in the chromatogram. But with MS and MS/MS detection sometimes many problems occur through heavy matrix effects. Many substances cannot be detected with MS/MS because of selectivity reasons, but on the other hand they influence the ionization of the analytes and internal standards (the compensation is actually what you want).

In the following there is one more remark on RP materials for precolumn clean up that do not need a conditioning step with organic solvents but can be used directly. These materials (e.g., ENV+) also can retard relatively hydrophilic molecules: Although these materials have a good retention for hydrophilic molecules surprisingly salts and most proteins from the plasma matrix can be eliminated through the clean-up step often using water only.

3.3.2.2
Mixed-Mode Phases

During the last couple of years two more phases with two separation modes in one column were successfully introduced in our lab: one is the RP effect through lipophilic behavior, the other is an ionic-exchange effect with weak and strong cationic and anionic effects. As many analytes are lipophilic acids or bases, this combination could be really helpful. Some of these phases are able to bind phospholipids strongly that often cause matrix effects with MS and MS/MS detection. This binding could be so strong that they do not elute even under elution conditions for the analytes.

3.3.2.3
Ion-Exchange Phases
The separation power of these phases exists exclusively on the reversible binding of ionic parts of the analyte. For specific sample preparation in the ultratrace range – naturally with fitting ionic analytes – these phases sometimes can help to solve a determination problem.

3.3.2.4
Special Phases

There are special phases on the market, for their use you have to read the instruction carefully. Let us look into one specific example to see how selective such a phase could be. It is worse by having a big sample preparation problem to search for such an option: Boronic acid columns! You may get them only from very specific distributors. Nowadays, with such a sensitive and selective detection as MS/MS it is not common to solve analytical problems in such an unusual way for clean up.

This phase catches very selectively all 1,2-diphenols under slightly basic conditions. On the other hand, all monophenols or 1,3- and 1,4-diphenols are not retained! Some years ago – it might be possible today too – this type of column was used for clean up during the determination of catecholamines. But also different phytoactive substances like catechins or gallyl molecules can be retained strongly with it.

4
HPLC Separation

The HPLC separation of diverse analytes, whether drugs, metabolites of drugs, naturally occurring substances or endogenous substances has changed dramatically since the start of (true) HPLC around 1970–1975. At that time the starting point was the experience of thin-layer chromatography (TLC) on pure silica gel. Roughly speaking, these silica materials were put under high pressure into columns for HPLC separation. It was a problem in the bioanalytical field from the beginning that analytes from hydrophilic matrices (like plasma or urine) were injected onto columns that had hexane or a mixture of hexane with small parts of some alcohols as mobile phases (at that time so-called "normal phase"). Trace amounts of water in the sample extract, which means also in the injection solution, changes not only the separation (peak elution order and retention time) in the given chromatogram but also destabilizes the chromatographic system sometimes for a couple of analytical runs. Within a short time separation on "reversed phases" (opposite to "normal phase") were developed. These – by their name – show a contrary behavior to the silica normal phases. These phases consisting of reversed "polarity" are much easier to handle. Furthermore, they covered the whole spectrum of most organic molecules, especially small molecules between 100–2000 Dalton. Extremely hydrophilic substances like sugars or large molecules like proteins could not be analyzed on RP at that time. As a good option for sugar separation amino phases turned out to be a good separation method with about 70% acetonitrile/30% water as mobile phase. As RP phases with lower silanol activity and larger pore size were developed the determination of proteins has become more and more possible. The hydrophilic bases and acids remained a problem with almost no retention on RP. By using almost only water as mobile phase (up to about 5% methanol or acetonitrile) the C8 and C18 tentacles broke down within a short time, resulting in almost no retention. With the upcoming development of ion-pairing chromatography (especially different long-chain sulfonic acids for bases and lipophilic ammonium substances like tetramethyl – or even better tetrabutylammonium hydroxide) a good retention on RP phases could also be obtained for highly hydrophilic acids and bases. As all these ion-pairing substances had low UV absorption (especially using >230 nm) they fitted perfectly. In our lab, we often used methanesulfonic acid or even perchloric acid to get a good peak shape (without tailing) for hydrophilic and also lipophilic bases. This tailing as a result of silanol activity remaining was

HPLC Methods for Clinical Pharmaceutical Analysis, First Edition. Hermann Mascher
© 2012 Wiley-VCH Verlag GmbH & Co. KGaA. Published 2012 by Wiley-VCH Verlag GmbH & Co. KGaA.

a terrible side effect in those times. After introduction of mass spectrometry as an HPLC detector the possibility of using ion-pairing reagents was strongly reduced because most of these ion-pairing substances are almost impossible to evaporate and this is why the MS instruments become blind, meaning very insensitive. As ion-pairing substances for bases by using MS detection only TFA (this results in almost no modified lipophilic behavior) and NFPA (nonafluoro pentanoic acid) can be used. Sometimes, the ionization is partially suppressed and the concentrations of these substances in the mobile phase should be low (<2–10 mM). As the column-producing companies have nowadays RP materials of only little remaining silanol activity most basic substances have almost no tailing any longer. Another way could be to make chromatographic separations with undissociated bases using a pH of 10–11. There are some RP materials on the market where you can use 10–20 mM ammonium carbaminate for days in water–methanol or -acetonitrile mixture without any problem. This type of separation for similar bases can result in wonderfully separated peaks – better than under acidic conditions. Sometimes, by using MS detection this could be helpful as you can detect your molecules with positive and/or negative ionization.

An HPLC system consists of one or more pumps, a column (sometimes with a protecting precolumn) and a detector. For determination of an analyte you have to consider for each case the options and limits of your own separation system.

4.1
HPLC Pumps

For simple separations an isocratic separation system (this means with one pump and no gradient mixer) is usually fitting . Sometimes, a gradient system is used isocratically only because the total analysis time per injection is much faster without backflush isocratically only and by using UV detectors with low wavelength (<230 nm) no baseline drift occurs.

Most low pressure gradient HPLC systems nowadays work with the usage of only one pump while high pressure gradient systems need as many pumps as there are different elution solvents.

For further details see Section 4.4.1.1.

A few further remarks on the combination of gradient elution with certain detectors:

- ECD (electrochemical detector) and gradient elution presumably never go well together, except you work at high analyte concentrations. Therefore, it does not matter, whether you work with glassy-carbon electrodes or with porous graphite electrodes.

- With a UV detector it depends very much on wavelength and on the means of processing whether a compound determination with gradient elution is successful. Not only can the organic share in the means of dissolving (especially methanol at low wavelengths) cause difficulties due to the fact that the baseline drifts away, the buffers in the watery share can also be problematic. Gradients with TFA and acetonitrile for peptides at 215–220 nm always show drifting baselines due to TFA or unclean acetonitrile.

- However, combinations of gradient elution with MS or MS/MS can also cause problems. Occasionally, the composition of the mobile phase – at the very moment of elution - experiences a labile phase. Little deviations of the retention time may be of great importance for ionization. Thus, an either bigger or smaller share of the organic solvent may exert a big influence, if the temperature for ionization is at a critical point. As a general rule it is recommendable to keep the buffers/acids/bases contents in watery and organic phase alike.
- You have to consider that buffers/acids/bases in organic solvents occasionally quickly reach their dissolving limits. With molecules that split off water during ionization you have to be especially careful: in this case it is recommendable to run isocratically and not by using a gradient and use ESI rather than APCI if ionization yield is good enough.

4.1.2
UPLC (Ultraperformance Liquid Chromatography)

On account of the introduction of UPLC (these are pumps for pressures of up to 1200 bar), today it is possible either to run faster chromatography or to apply even smaller particles in HPLC columns (from about 1 to 3 µm). When you use UPLC you have, however, to pay attention to apply a certain general setup, otherwise you have to face the fact that there is no better separation than with conventional HPLC.

- It is important among other facts, that you use only capillaries with a thin inner diameter, as recommended by the manufacturer of the device.
- Generally speaking, column-filling materials of which the particle diameter is too large should not be used, because by using too high flow rates you are already far beyond the optimum of the Van-Deemter curve (too high flows cannot be regulated with all the setups of most manufacturers).
- The smallest particle sizes (1–2 µm) bear the immense disadvantage including entrance and exit frits of the columns, that you should not inject into them samples containing cloudy or suspended solutions. This is opposite to the practical experience in former times with 5-µm particles and column dimensions of 125 mm x 3–4 mm id, as the pressure resistance will become too strong. Please pay attention to this fact otherwise you could lose your expensive ULPC column.

4.2
Degasser

Irrespective of which HPLC system you apply, the mobile phase should be either purged with helium or should run through a vacuum degasser (ultrasound only plays a subordinate role). Apart from that you meet big problems especially with gradient elution, that is to say the following ones:

- You find more deviations in pressure (thus, with UV detection also an irregular base line appears, with electrochemical detection an extremely noisy baseline will occur).
- With UV detection with wavelength under 230 nm small bubbles appear again and again as big peaks in the chromatogram (and often cannot be discerned as bubbles).
- With MS and MS/MS detection (maybe also with electrospray ionization) the subtle spray tears off and many zigzag lines appear in the chromatogram.
- Retention time varies from injection to injection.

4.3
Injector

Most of the autosamplers nowadays are optimized on

- very reproducible injection volumes;
- low carryover.

With respect to the fact that you already run one or various systems in the laboratory (thus you cannot buy the optimum autosampler), we give you some more details on a few mistakes that we came across while doing our research work at our laboratory over a few decades.

4.3.1
Big Carryover

One of the reasons why carryover takes place is that you injected an extremely concentrated solution during HPLC method set up shortly before, thus the manufacturer of the device is not to be blamed. Another reason could be an injection loop that has been used for a long time. Therefore, the inner surface is already corroded and thus has developed a "mega-memory".

Also, some of the rotors are either noticeably scratched or dirty and these need to be cleaned or replaced from time to time.

Furthermore, the composition of the injector-rinsing solution is immensely important: Lipophilic substances or lipophilic matrices contained in the injection solution also require a lipophilic injector rinsing solution. With bases also certain acids usually are applied in order to easily solve the remainder as salts. The other way round one tends to use also basic additives with acids. A certain solvent applied up to a certain percentage has proved to be useful: DMSO. It is able to solve many substances. However, you must observe that you work with an air cushion, because otherwise small quantities of DMSO get injected into the column, which can suppress the MS ionization.

We frequently use our own special composition with the injector's rinsing solution for each research problem.

4.3.2
Deviations of Injection Volumes

Many autosamplers apply vacuum for induction, if the induction steps are carried through too rapidly, air bubbles may be caused – especially with rinsing steps of big volumes. These air bubbles may lead to relatively big variations of the injection volume, which makes it necessary to use an internal standard.

Also, minor leaks in the induction system may cause air bubbles.

4.4
HPLC Columns

You could write 10 books on this subject, without referring to all aspects of this complex, extensive topic. One thing you may mention in advance: if one kind of column has proved to be successful (stationary phase, column dimensions, occasionally also a certain manufacturer) when applied in your laboratory, then you should not change it, no matter what various manufactures promise during their new stationary phase development. Possibly you would lose in this way part of the feeling that you had already acquired for the use of mobile phases and their composition.

However, if you do change to a new separation system, – though it may be a better one – the former validation is obsolete if it was in the GLP/GMP area.

Even if you as a user do not work within the regulated area (GLP, GMP, maybe ISO) you have turned all findings upside down except for stability examinations, because you have changed the stationary phase.

You may of course try a new type of column that may lead to an excellent surprise. However, to use this column you should wait until the next research topic or development of the method.

4.4.1
Reversed Phase – Phases with Reversed Polarity

Simple reversed phases suffice to use with many applications.

These "reversed phase" columns are offered in great variety. Important criteria for your choice may be:

a.) If it is an exotic column or an exotic manufacturer, then the next column batch you order will surely be problematic in terms of comparability with the old one.

b.) Chose the proper column dimension for your kind of research work. So, for instance, a 250 x 4 mm column with 5-μm material is oversized when you have a simple separation. With our application experience (even during that period while we still used to work often with UV and fluorescence detection) 125 x 3 mm columns with 5-μm material are our standard. These columns are robust and are clearly more economical in their consumption of acetonitrile than 4 to 4.6-mm-id columns. They don't get clogged up as easy as 2-mm columns (especially if you use 3-μm particles) and their costs are reasonable. For MS applications we often use 50 x 2 mm columns (often filled with 3-μm particles) or 150 x 2 mm columns for ambitious applications. These columns have relatively short dead times and good separations by using flows of 0.5–0.8 ml/min that can be done within 3–5 min (k' values of 3–6).

c.) Try to use two to four quite different materials in your lab. So, if you have separation problems you can easily switch to another type with quite different characteristics.

d.) One of these columns should be a so-called "AQ" column. This means a column where you can use 100% watery mobile phases without collapsing lipophilic tentacles (trade names could be AQ, Hydrosphere, Pyramid, Polar-RP or Aqua).

More details concerning RP phases will not be mentioned here.

4.4.1.1
Mobile Phases for Reversed-Phase Separations

Regarding mobile phases for separation on RP materials there is much more to be said. Furthermore, it does not highly influence the selling activities of the different producers of HPLC columns.

a.) If it is not really important for your separation whether you use acetonitrile or methanol then use acetonitrile. Concerning efficiency this is the only fitting answer. Looking at the pressure profile you can see that having the same separation power chromatography with acetonitrile is much faster than with methanol. Using methanol you will have many problems with your chromatographic baseline in noise and drift if you use the low-UV range between 195 to 240 nm. It is not so bad under isocratic conditions but really worse with gradient elution. Arguments against acetonitrile are some toxicity aspects and – looking back – some supply shortages. Certainly, you will take methanol instead of acetonitrile for specific separations where a substance group has a higher selectivity in separation with methanol. Sometimes, by using MS detection much higher signal abundance can occur. Please take into account that you need a good reason to use room temperature for separation. Usually, using 40–60 °C makes everything smoother: the backpressure is much lower and enables higher flow rates. The separa-

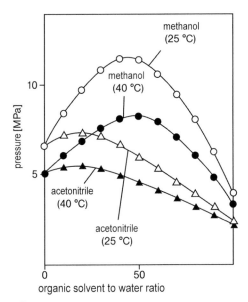

Pressure Profile

tion power is higher. Higher flow rates especially with gradient separation make the backflush much faster and shorter.

b.) Keep in mind that only "real" neutral analytes need no buffered system (buffers, bases, acids). Sometimes, separation for your analyte without a bad matrix could be stable (retention time, peak shape) without stabilizing the mobile phase with buffers. But analyzing your dirty samples without an excellent clean-up step before HPLC separation can end in a fiasco. It could be that you can't find your analyte in your sample any longer because of shifting retention time for each new injection.

c.) For analyzing basic analytes using a counterion is a good decision. Counter ions could be trifluoroacetic acid (good for peak shape) or nonafluoro pentanoic acid (for good peak shape and better retention). Alternatively, you could use an alkaline mobile phase and chromatography with the nondissociated molecule. For this kind of separation you usually need a higher percentage of organic solvent for the same retention time as for using ionpair chromatography under acidic conditions. Chromatography under alkaline conditions often has a completely different selectivity than with ion-pair chromatography.

d.) In former times, we used substance like tetrabutylammoniumhydroxide for ion-pair chromatography for acids or even more lipophilic compounds with UV or fluorescence detection. Nowadays, we usually do not use this technique any longer when using MS detection. With this technique it can be possible to put the analyte as an ion pair into a "clean" chromatographic window aside from endogenous substances. This "fine tuning" concerning c.) and d.) to find this window is easy to do. For a basic analyte a minor variation of the concentration of the acidic ion-pairing reagent a strong influence on retention time is the result: Taking 10 mM pentane sulfonic acid instead of 8 mM the peak will elute considerably later. By contrast, taking 6 mM the peak will elute clearly sooner. This effect can be gained in the same way with octane sulfonic acid or decane sulfonic acid. But keep in mind that for longer sulfonic acids you need some percentage of organic solvent to solve the ion-pairing reagent. Sometimes, it is also necessary to put this ion-pairing reagent into the injection solution too. So, the analyte will start immediately as the complete ion pair.

e.) Please be careful when using volatile constituents of the mobile phase like ammonia. All fittings have to be tight. For being certain take a new mobile phase after 2–3 days usage of the mobile phase (because of partly evaporation of the ammonia).

f.) Please take into account that by using isocratic conditions it could happen that you see broad peaks in later chromatograms. These are late-eluting substances from injections, e.g., one hour before – this means 15 injections before. Then, you will see those broad peaks in each subsequent chromatogram. This is a bad effect especially when you have only an isocratic pump. Otherwise there is an easy solution: Make a step gradient to 100% organic after, e.g., each 15th injection for cleaning the column from late-eluting peaks.

4.4.2
Normal Phase – Phases with "Normal" Polarity

Bare silica gel or silica gel with free endstanding NH_2, OH or CN groups can be used under so-called normal-phase conditions. This means with very lipophilic solvents. As mentioned, this is a very potent separation technique especially with minor variations in functional groups. But be careful as small quantities of water, e.g., in the injection solution can disturb the whole chromatogram, and sometimes also the following chromatograms. In the respective literature you can find impressive separations under normal-phase conditions.

Some of these stationary phases can also be used for another kind of separation:

- *NH_2 columns*: for determination of sugars with a good separation with about 70% acetonitrile/30% water (more acetonitrile means longer retention times). For these separations the ion-exchange effect is used.
- *CN columns*: These columns often possess a short aliphatic chain prior to the endstanding CN. So these columns are using the reversed-phase effect with little retention (short aliphatic chain) and a slightly different selectivity because of CN.
- *Bare-silica columns*: Roughly 20 years ago some publications were issued with extremely selective determination of basic compounds on bare silica using quasi-RP conditions concerning mobile phase[6,7].

---

6 Anal Chem 54 (1982) 442-447 "Separation of Organic Amine Compounds on Silica Gel with Reversed-Phase Eluents"

7 Chromatographia 24 (1987) 713-719 "Use of Unmodified Silica with Buffered Aqueous Mobile Phase mixtures for Selective Chromatography of Basic Drugs"

Finally, a Si–OH surface works like a weak cationic exchanger with 80–95% acetonitrile or methanol and 5–20% water with 5–20 mM acids – often strong acids like TFA, TCA or methane sulfonic acid. The separation power for primary and secondary amines is impressive. But be careful not to use MS or MS/MS as the means of detection. Metabolites of basic parent drugs often have the same basic activity as the parent compound and therefore the same retention time (e.g., hydroxylation, demethylation far from amino group). By using UV or fluorescence detection no differentiation is possible. Therefore, although such an extreme selectivity in separation is achieved to other amines no selectivity between parent substance and metabolites is given.

4.4.3
Ion Exchanger

For specific questions concerning selective separation strong ion exchangers (anionic or cationic) are a good, or the only, alternative to RP. As for elution on these systems a salt gradient is often helpful, of course using the correct salts! MS detection can only seldom be used.

4.4.4
Carbon Columns (Porous Graphite Columns)

In a simple way you could say this is a RP column but this is not strictly true. This RP effect needs a much higher percentage of organic modifier for elution than C18. Besides, these columns favour aromates in two ways – stronger retention in relation to aliphates and more selectivity among aromates. For some applications with MS detection a direct comparison of porous graphite columns and C18 showed the following: For hydrophilic analytes that have on C18 almost no retention, although 100% water have a good retention with 10–20% organic modifier in water on carbon[8]. By using less water the sensitivity with MS detection is sometimes strongly increased.

4.4.5
Enantiomeric Separation

Using enantioselective separation needs a good portion of courage and experimental spirit. Courage because of the high price of these columns and their fragile behavior if using the wrong conditions. Experimental spirit because of the necessity of chromatographic feeling for tuning the mobile phase to get separation and also enough separation of the enantiomers within a relatively short time. Two things are not desired: One run of 30 min means only about 20–40 injections per day and broad peaks after such a long retention means low detection sensitivity.

An enantioselective column has quite different separation mechanisms from RP separation and besides, each type has its own special behavior. This means for practical reasons that you have to decide for only few types of columns. These columns usually should be useable under RP conditions (water/organic). Otherwise, the sample preparation and especially the injection solution must be changed and adopted dramatically. With many these enantiomeric column types the flow, respectively, the pressure must be observed carefully. The pressure limit of most of these columns is only 25–40% of RP columns. Please use the pressure-limitation settings of your HPLC system! The next step is to read the instructions of the producer for this column – there are often forbidden solvents or salts and suggestions for getting separation of enantiomers. This could be pH changes, buffer concentrations or even lowering the column temperature (pressure goes up!) for better separations. If you can't find your analyte in the separation list than look for structurally similar substances (look at the structures: sometimes in bulky and sometimes in functional groups). In our opinion cyclodextrine columns need much patience and have good separation often after very long run times. We use modified cyclodextrine columns sometimes for isomeric separations with much success. AGP, CBH or BAS columns – this means columns with protein coverage – are really useful for many racemates. But higher percentage of organic solvents can kill your column! Also important when using these columns are relatively clean extracts of plasma or urine samples – cleaning these columns often ends without success or with destroying the columns. Many enantiomeric separation columns can be cleaned with a higher percentage of organic solvents or high buffer concentrations or specific acids or bases. For protein-covered columns all these cleaning procedures are prohibited because of denaturizing the proteins on the surface.

In the meantime there are many different stationary phases for enantiomeric separation on the market. In our lab we have 5 to 7 different column types with different separation mechanisms in use. We know these columns well and for a new separation we check one column after the other until we have success. For election of the column types we always carefully read the suggestions of the producer for separation mechanism. Please always use a racemic mixture for your tests! A brilliant peak for one enantiomer is far too little information to recognize an enantiomeric separation! We suggest the best (= cleanest) extraction procedures when using enantiomeric separation, which is LLE for lipophilic molecules. A good guidance for enantiomeric separation can be found in the book edited by Wolfgang Lindner[9].

---

9   See list of books: Lindner

5
Detection

Many thoughts and suggestions for detection can be found in Section 2.3. During many years we developed methods for determination of hundreds of drugs, metabolites or endogenous substances in plasma, urine or tissue. For all these determinations our first and central thought was detection. As often stressed, we always start all the considerations with the necessary LLOQ (determination limit) and the detectors we have at hand. What is the LOD (limit of detection) for a given substance? First, we do not consider the selectivity concerning possible interfering matrix substances. This is our second step – sample preparation and analyte enrichment. All these considerations may later then exclude even detectors where the LOD was suitable.

Useful tip 7: Complexity of analysis

At this point some general considerations concerning complexity of analytical questions in this clinical/pharmaceutical area should be mentioned:

a.) In the pharmaceutical area often preclinical and clinical samples (mostly plasma or serum) have to be analyzed. These samples were drawn after application of usually one drug. This application can be done in different ways: orally – often then with metabolites after first-pass effect; "iv" with relatively high levels of parent drug or "im" or "sc" with variable concentrations of parent drug. As a consequence, only problems with matrix substances (= endogenous substances) and resulting metabolites can occur – no more and no less. To describe the pharmacokinetic profile for a given drug or its metabolites rather low determination levels should be gained. At least 5% of C_{max} but much better <1% of C_{max} to see clearly the terminal elimination phase.

b.) By contrast, the question for "clinical samples" that often deal with therapy control or compliance only. Therefore, the question for extremely low determination limits is not as important as with pharmacokinetic profiles. On the other hand often multimorbide patients get many different drugs at the same time. So these other drugs and sometimes 50–100 different important metabolites can occur and can disturb the determination of a given drug. Therefore, the determination must be sometimes much more selective but less sensitive. As a consequence, the detection should be more selective (less UV

---

but more MS detection) and the sample clean up must be rather selective. Sometimes, add on of drugs to the extract can help to prove that the huge peak is surely the given drug.

5.1
Detection in the Pharmaceutical/Bioanalytical Area

After oral application of more than 100 mg per person and little "first-pass effect" you can be sure to get a maximum plasma concentration of more than 1–10 µg/ml.

In this concentration range UV or fluorescence detection would give the most precise results (precision, accuracy), as long as the analytes have a good response with these detectors. But FDA and EMA have the tendency in recent years to force the use of MS/MS detection. MS can suffer from strong matrix effects. This tendency partly comes from the fact that getting low determination limits with UV or fluorescence detection sometimes derivatization or online-sample enrichment is needed, which is not everyone's business. On the other hand MS/MS systems are rather expensive but can be easily bought. The so-called "matrix effects" can be met with isotopic-labelled internal standards. Besides this, these isotopic labelled internal standards are rather expensive and have to be synthesized on your own behalf. The second problem is therefore that usually only the parent drug then possesses an internal standard. For the metabolites of this parent drug this internal standard may not be suitable at all. Further, metabolites have other retention times and not seldom different ionization behavior. Therefore, for the parent drug isotopic labelled internal standard sometimes or often will not be fitting for their determination. For practical reasons at the beginning of a new drug development in the preclinical area the parent isotopic labelled internal standard is often used for quantification of the metabolites, too. In such cases the internal standard only eliminates dilution effects (sample clean up or HPLC injection faults) but cannot eliminate specific matrix effects. Furthermore, specific sample clean-up steps for metabolites opposite to the parent drug makes the internal standard useless (different recovery rates for example).

This tendency of authorities is a pity because these good and stable detectors that are also cheap are being more and more withdrawn from circulation, without being necessary. Imaginative analysts are not as easily purchasable as expensive instruments.

5.2
Detection in the "Clinical Area" (Therapy Control/Compliance)

For this area the tendency should go to more selective detectors. Seldom one can achieve such a selective sample clean up for specific substance classes that the selectivity of detection is not so important. The demand for high precision and accuracy is not so important in this area with few exceptions. But the selectivity is an important point by analyzing samples of many different patients – often only one sample per person. In the pharmaceutical area with preclinical and clinical samples (mostly phase 1–3) often a predose sample exists and therefore a good control of selectivity is given.

But the reality is often not the same. Labs in the pharmaceutical area often have much more money than labs in the clinical area for buying rather expensive instruments.

6
Chemical Derivatization for Detection Enhancement

There is always the question – which detectors do I have in my lab? If I have MS/ MS systems rather seldom is there the necessity for detection enhancement.

Most colleagues have bad associations with derivatization. This reminds them of chemical synthesis labs with bad reaction percentage and many side products. They think they will get more problems rather than solving problems.

Potential objections:
- If you want to use "precolumn" derivatization the reaction has to be quantitative. But is that the reality? Our experience as chemists is quite the opposite! 5–20% total effect after some synthesis steps are taken within a few weeks. How can I find a way for quantitative derivatization? Moreover, many side products occur during synthesis and they can disturb as unwanted peaks in the chromatogram. Furthermore, you can't take plasma or urine as it is and make that derivatization. No, I do not want or dare such a derivatization – some might think!
- For a so-called "postcolumn" derivatization there are more objections to be made. On the one hand, more pumps and ancillary equipment are needed because of pumping a reagent after HPLC separation to the eluent in front of the detector. On the other hand, the reaction has to be within seconds – does that ever occur? And also the demand for quantitative reaction exists. Furthermore, you have no option to choose the matrix for reaction – it is the eluting mobile phase. In addition, the peaks will become wider (this is only an argument of real experts!).

Summary: Derivatization is not a tool I want to use!

Useful tip 8: Our incentives for derivatization
Under which circumstances do we use derivatization?
The answer is never when solutions are possible without derivatization. The steep development of detectors that are universal, selective and sensitive is still in progress. So with MS/MS systems derivatization is needed only for very delicate demands. One example should exemplarily demonstrate also for MS/MS systems that derivatization could be really necessary. If you have to determine estrogens in plasma down to 1 pg/ml – especially for EE2 (ethinylestradiol) – you have almost

HPLC Methods for Clinical Pharmaceutical Analysis, First Edition. Hermann Mascher
© 2012 Wiley-VCH Verlag GmbH & Co. KGaA. Published 2012 by Wiley-VCH Verlag GmbH & Co. KGaA.

no chance without derivatization even with the most sensitive MS/MS systems on the market. Without derivatization only about 1–3 pg can be seen as a peak for EE2 – also with sensitive systems. You may think that variation of mobile phases could solve the problem or changing the ion source or the temperature of the ion source – nothing is helpful! There is no normal way to solve this sensitivity problem.

Yes, there are many substances where you can see on an API5000 even 5–20 fg per peak but not for estrogens.

The only solution is "derivatization" of the phenolic group. In the literature you can find much about derivatization of phenols – often with dansylchloride. If you are a newcomer in this field – don't start with such a difficult topic. Use another topic for derivatization with concentrations much higher.

Thinking of derivatization – also a derivatization opponent does not know a better way than derivatization: Amino acids! For these molecules the starting point was ninhydrine derivatization roughly 30 years ago. Then derivatization of the primary and partly secondary amines occurred (a long time before MS/MS was an instrument in many labs) with Fluram, OPA, NDA, AQC ... pre- and postcolumn. That was a triumphal procession through the HPLC analytics!

If you are a newcomer in the field of derivatization just try an amino acid derivatization with two or three different amino acids for training reasons. Start with precolumn – derivatization with Fluram or OPA. Derivatization with Fluram can be done by mixing the sample with buffer (pH 8–9) and afterwards with a solution of Fluram in acetone or acetonitrile. The reaction is completed within seconds. Fluram must be solved and stored in acetone or acetonitrile. Water will destroy Fluram immediately. For derivatization with OPA (ortho-Phthalaldehyde) you need a buffer of around pH 9 and a thiol. Some different thiols are described in the literature. Usually, mercaptoethanol is used but the resulting derivates are not particularly stable. But for training reasons this is good enough. Thiol and OPA will be mixed with buffer pH 9 and pumped postcolumn through a T-junction to the eluate. The combined phases need some seconds for reaction in a Teflon coil before fluorescence detection. For OPA as well as for thiol you can use 10 mg/l. In the same way you use precolumn derivatization: Sample plus buffer (including OPA and a thiol) vortexing and after around 5 min you can inject the sample. If you use a UV detector you can also see the derivatives of amino acids with Fluram or OPA well, but the access of the derivatization agent will disturb in the chromatogram. By using fluorescence detection with the appropriate excitation and emission wavelength you will really see a miracle: extreme sensitivity and selectivity of the derivatized amino acids. If you now are a little familiar with the topic "derivatization" then I will suggest a specific book to you which from the practical point of view is the best book for derivatization in my opinion: "Detection – Oriented Derivatization Techniques in Liquid Chromatography" by Lingeman and Underberg (see Table 11.36), especially Chapter 7 for UV detection and Chapter 9 for fluorescence detection is a rich source. Also, in the Appendix some books are cited – there are only a few practical books on the market concerning this topic. Also, some review articles can be helpful[10][11][12][13]. From our point of view you will have the biggest profit from derivatization in combination with fluorescence detection. Fluorescence often means: No peak before reaction and sometimes a huge peak after reaction. By using UV detection not seldom some side reaction products can be seen in the chromatogram and a huge peak from the excessive derivatization agent. If you look carefully into the literature you will not only find a few derivatizations for HPLC. By contrast to HPLC, GC often needs derivatization for getting volatile analytes. Especially in the area of steroids or sugars you need two or three different derivatization agents and steps for getting a volatile substance by covering different functional groups. HPLC usually needs derivatization for detection reasons. Each derivatization step makes the analytes more similar and therefore worse to separate within one structural group. This is not the effect we usually want. Sometimes, derivatization for HPLC is necessary because of sensitive functional groups (e.g., thiols). In the dawn of HPLC the stationary phases were not so inert. Therefore, sometimes derivatization of especially basic groups was helpful for getting sharper peaks without tailing. So, especially in trace-level detection strong tailing peaks gave only a tailing hill rather than a peak.

10 J Chrom Sci 17 (1979) 147-151 "Fluorimetric Derivatization in HPLC"
11 Analyst 109 (1984) 47/01 „Fluorigenic Reagents for Primary and Secondary Amines and Thiols in HPLC"
12 J Chrom 141 (1977) 107-119 „Pre-column Derivatisation in HPLC"
13 J Chrom Sci 17 (1979) 160-167 „UV-Visible Absorption Derivatization in Liquid Chromatography"

7
Validation Concepts

7.1
Introduction

From my point of view one mistake should not be made by an analyst working in the bioanalytical field: "To put the blame on someone else!" Why do I say this when dealing with the topic of validation concepts?

Working in the area of clinical/pharmaceutical analysis the topic is so complicated that an analyst could be slightly overburdened. We work in an area where the analytical instruments become more complex each year. Therefore, working with such machines is challenging. Besides the matrix plasma, urine and tissue are not the easiest ones. This extreme complexity with all the limitations can easily lead to many problems. In this despair your reaction could be that the clinician who has generated the sample will be blamed by you. But in reality the sample was interchanged or the autosampler was blocked or leaky or hasn't injected any sample. The mentioned failures are relatively easy to find. But often there are mistakes that are terrible to clear up. In such cases you can see different reactions from colleagues. The average analyst briefly checks the system, does not find the reason for dysfunction and then complains about his boss who developed this method. Or he blames the clinician who has delivered a wrong sample. I try to tell my staff that they can only quarrel with others, after checking everything really carefully.

Why all these statements when talking about validation? Because it is necessary during method development and method validation to observe all these facts. The analyst must keep in mind all the facts about the substance (who is able to do that?), the limit of determination, the matrix, the possible substances disturbing the determination, the possible instabilities, the potential variations concerning different matrix constituents from different patients having different sicknesses ... So, he has to recognize which of these points can be dangerous for correct results. Exactly these points must be examined carefully during method development and method validation. If you take the FDA guideline in hand concerning validation of bioanalytical methods then you can see that they had many of these dangers in mind – but by far not all! If you want to be on the "safe" side then you had to analyze at least 400–500 samples during method validation. This is far from any reality. So as a "doubtful" analyst you had to find a practical way to do this – otherwise no one will pay for all the effort! Each analyst had to gain compromises between his aware-

HPLC Methods for Clinical Pharmaceutical Analysis, First Edition. Hermann Mascher.
© 2012 Wiley-VCH Verlag GmbH & Co. KGaA. Published 2012 by Wiley-VCH Verlag GmbH & Co. KGaA.

ness, the imagination of his boss, the ambiguous stated requests of the authorities and the disposable working time. But one thing you must keep in mind: The first and most important thing is "correct results" and not "fully validated methods". So, in-house we often had discussions when – by using a validated method in routine – some new questions arose (we mostly work in the GMP/GLP regulated area beside feasibility studies or some early preclinical studies). Could we clear up those questions or ideas in a GLP sequence calling them "test samples"? Our QS department forces us to define those "test samples" more clearly beforehand because of possible questions arising during audits of sponsors or authorities. It is a shame that we can not make a definite decision. One thing I can surely say is that an extremely good analyst can easily destroy a validated method because not everything can be tested. In the old Spanish literature of the 16th century there is a poem of Calderon de la Barca: "the judge of Zalamea" where you can read: "The one who really meets the point may be wrong in little things!"

Let us make this a mission statement that we don't miss the point:
"Correct results"
A simple example could be a LLE with extraction under basic conditions and following re-extraction from organic solvent und acidic conditions: Tell me more about stability after a basic buffer was added to the plasma?
How long are the analyte and the internal standard stable?
   This is relatively easy to answer. But tell me something about degradation products from the matrix under basic conditions that are able to disturb the determination of the analyte? These degradation products are generated more and more within a certain time. Is it necessary to know those substances from the chemical structure? How is it with plasma from different kind of patients: lipemic, high bilirubin, from liver cirrhotic patients, from renal disease suffering patients ... . How many different matrices have to be checked? Using how many spiked samples in which concentration range? Is it necessary to keep the samples at around 20 °C or is it allowed in summer time to have 30 °C? After first extraction under basic conditions are the analyte and the internal standard stable? Stable before or after centrifugation? For 10 min or also 2 or 3 h? Is this conclusion correct – or do it as fast as possible? Does this mean also fast re-extraction into acidic buffer? We saw that the analyte is sold as a salt – a hydrochloride. Is 0.1 M hydrochloric acid OK or not for re-extraction?

   You see many questions – and you have to find a practical way!

   Do you understand the problem? Each step has to be thought over carefully and compared with your own experience (which is growing each day!).

   What we want – as mentioned before – are correct results, and the validation should help to reach this goal!

7.2
Realization of the FDA Guideline

In analytics under "method validation" one understands the official and documented confirmation, that an analytical method is adequate for the specific application and that it meets the demands in question. A validated bioanalytical method has to lead to correct, reproducible data, so that a valid interpretation of the studies is possible. In May 2001 the CDER (Centre for Drug Evaluation and Research) of FDA published its official guidelines for bioanalytical methods.

This document with the title "guidance for Industry, Bioanalytical Method Validation" contains general recommendations for the validation of bioanalytical methods, which are used in clinical human studies and that require a pharmacokinetic evaluation. The guidelines are applied, however, with toxicological and preclinical studies too, which are carried out on animals.

Example for a task: determination of an analyte in a predefined matrix in the requested concentration range. The bottom edge, which means the edge of determination, will be defined by the rule 5% from C_{max}, the upper edge by the linearity of the detector and the maximum therapeutic levels in, e.g., plasma.

Usually, on top of a validation method there is the forming of a validation plan, in which the necessary validation steps as well as the criteria of acceptance are mentioned. This is the condition for an accomplishment according to GLP.

First, it is necessary to fix a mathematical model for calibration (response versus concentration of the analyte). For calibration of the measurement instrument one needs self-produced calibration samples with a determined content. Such samples are usually made through dilution in mixtures of an organic solvent and water. The choice of the model is essential, as most of the bioanalytical methods have to meet a large concentration range. Unweighted "least square" linear regressions only show acceptable results for a small concentration range. Under larger concentration ranges the higher standards influence the variance of the smaller standards and so this influences the result. Therefore, an evaluation of 1/conc., respectively, 1/conc. 2 is used.

a.) How many standards (calibration samples) are necessary for a calibration curve? This depends on the calibration range, but the FDA claims a minimum of 6 to 8 standards, which cover the total level of LLOQ up to ULOQ. In addition to that, 2 void matrix samples, one with and one without internal standard, are necessary. All samples, which should be quantified in a measurement group together, have to be prepared within the same time.

b.) Dynamic calibration range
This depends on the acceptable criteria precision and accuracy. In the "consensus conference" of the FDA 1990 a range of ±15% was fixed, except with the LLOQ, where ±20% is allowed.

c.) Criteria for LLOQ
c1)   The analyte peak should at least be 5 times higher than the noise after injection of a void plasma sample.
c2)   The analyte peak has to be assignable and reproducible decisively with a precision of 20% and an accuracy of 80 to 120% related to

the theoretical value. At least 5 determinations have to be made with LLOQ in a block of analysis.

d.) Precision of the method

The precision has to be determined in three blocks of analysis as interbatch precision (= repeatability). In the end, the first block of analysis has to be repeated twice with new prepared analyze samples. As far as possible and acceptable it should be done on different days, with different analysts and different instruments. (This last claim doesn't sound irrational, but it is almost never completely viable!)

e.) Selectivity of the method

This only expresses that it must be possible to notice clearly the analyte in question under a big quantity of different substances (like from plasma and urine). Based on at least 6 different individual samples of the matrix without the addition of the analyte it is examined whether the identification of the analyte is disordered. In this respect it is irrelevant whether a nonselective UV detector or a sophisticated MS/MS system is used. This selectivity examination should be done referring to the verification range. One should be aware that many dimensions belong to selectivity – among others also matrix effects within MS/MS systems.

f.) Recovery

Finally, it is all about efficiency of the extraction of the analyte coming from the biological matrix. This is verified through comparison of the peak areas of the extracted samples with diluted stock solutions, whereby all dilution steps have to be considered. The recovery does not have to be 100% but consistent, precise and reproducible (including the internal standard). Recovery experiments should be done in three concentration steps (low, middle, high). We often use the 2.5 fold of the LLOQ, the median and around 80% of the highest standard concentration.

g.) Stability of the analyte

The stability of the analyte in the biological matrix during the sample processing and within the injection solvent is an important part of the total validation. In particular, the stability in the biological matrix as short-time stability at room temperature (e.g., 4 h) and three time-defined freeze-and–thaw cycles belong to the basic validation. At this moment, usually a LTS (= long-term stability examination) starts under for instance –20 °C or –70 °C for mostly 1–6 months. If samples out of clinical studies are on stock before being dealt in analytics, this time has to be covered in the validation. Further important stability examinations refer the stability of the stock solutions, the stability of the sample extraction (e. g. serum from whole blood) or even also the stability during the different steps of the sample preparation and in the injection solvent.

h.) Validation report

At the end of a validation one must create a report including all details of the sample preparation and sample measurement. It also contains all examination results regarding precision, linearity, accuracy, recovery, stability and selectivity. Then the method for analyzing the real samples is ready for use.

Regarding the real samples (respectively, even during the validation) all findings about metabolism (as long as existing) should be considered, so that metabolites are not included in the parent substance through unselective detection or inadequate chromatography.

Exploratory analysis is often qualified in quite another way, as the calibration range is partly unknown.

8
Practical Hints Concerning Stability, Destruction and Degradation Products

What could be dangerous for the stability of an analyte in plasma and urine?
  – Light, oxygen, temperature, enzymes, pH conditions, other matrix constituents.

Which analytes are especially endangered?

a.) Thiols (free SH-): They tend to build –S–S bonds with other thiols or to be oxidized in another way. Most of these reactions can be reversed by DTT (dithiothreitol) or other reduction agents like tributylphosphine (see Useful tip 9). Not always can the structure of the original molecule be gained with these reduction steps. For some reason also the question could be how many per cent of my analyte is oxidized or still in the matrix in reduced state. Then you need many tricks to get the right information.

b.) There are substances that are light sensitive in solution (for some substances you find some remarks in the literature), but most substances are not light sensitive, especially when you do not put your plasma or urine into direct sunlight (this shouldn't be done also because of temperature reasons). Some analytes –, e.g., Nifedipine and this substance group – are not stable under daylight or the usual light in labs. They will be destroyed within a short time. For practical reasons wrap all your centrifuge tubes and blood-drawing tubes with aluminum foil. Some colleagues used for those sensitive substances labs with a special light in dark labs[10].

c.) Enzyme activities – especially in plasma – are a big danger for some substances. In particular, peptides are destroyed sometimes within minutes. For peptides as analytes try a protease inhibitor – often it works and results in stable peptide concentrations for hours. Sometimes, a pH change is helpful – consider that enzymes are proteins with a specific isoelectric point.

d.) One point you should keep in mind: Plasma pH changes from pH 7.5 up to even 8.5 during storage at room temperature. This could have an influence on the stability of the analyte (seldom!) but also in recovery by using extraction with extremely pH-dependent molecules (pKa 7–9).

---

10 Chromatographia 25 (1988) 919-922 "HPLC-Determination of Nifedipine in Plasma on Normal Phase"

HPLC Methods for Clinical Pharmaceutical Analysis, First Edition. Hermann Mascher
© 2012 Wiley-VCH Verlag GmbH & Co. KGaA. Published 2012 by Wiley-VCH Verlag GmbH & Co. KGaA.

e.) What is true for thiols is also true for other easily oxidized molecules. Either determination of ascorbic acid from plasma or catecholamines – please check the literature for such labile compounds. There are some remarks for stabilization in the literature. For specific labile substances often analyzed, e.g., catecholamines or Levodopa, there are blood-drawing tubes on the market with stabilizing agents in the tube beforehand[11].

f.) Detailed remarks for best blood-drawing conditions, sample storage after plasma separation or fitting time points during the day (especially for circadian fluctuations during the day for hormones or other endogenous substances) please check the book written by Walter Guder[12] that we suggest explicitly.

g.) Concerning endogenous substances: Please keep in mind that such substances can be in the body under free and bound conditions in plasma or tissue. If you do not store your samples appropriately the "bound" substances can be set free and therefore manipulate the free level (also the most sensitive and selective method will give wrong results!).

h.) If you analyze urine samples then keep in mind that the whole sample must be homogeneous. With urine sometimes precipitations must be solved before taking an aliquot. Sometimes, 37 °C and/or ultrasonic bath are necessary to get there.

For long-time storage (LTS) of plasma or urine, <–20 °C usually is sufficient. But sometimes you may find for specific substances in combination with specific matrices remarks in the literature that instability was observed during LTS. We use three different conditions for LTS (>3–4 weeks): usually <–20 °C (range –20 to –30 °C), for sensitive substances <–70 °C or for extremely sensitive substances (only a very short thaw cycle) liquid nitrogen at –196 °C.

---

11 For example blood drawing devices for Catecholamines and Levodopa
12 See list of books: Guder

9
Metabolites

Metabolites can be extremely helpful or may lead to an analytical disaster. One remark in advance: Each enemy you know you can meet in a better way! And each metabolite is an enemy that disturbs or can disturb the determination of a given drug.

It will become a friend only under the condition that the metabolite is easier to be determined through much higher levels or better fitting structure. For example: higher levels and much longer plasma half-life (e.g., the parent drug with a half-life of minutes) therefore bioequivalence or compliance is much easier to determine. Often metabolites are more hydrophilic as the parent drug (demethylation or hydroxylation – phase 1 metabolites). Phase-2 metabolites (e.g., sulfation or glucuronization) are often too hydrophilic to be extracted with LLE, although the parent drug is rather lipophilic. Phase-2 metabolites are much more complicated to synthesize than phase-1 metabolites. So often no reference substance can be purchased. If you have to analyze a drug that is supposed to metabolite in different ways please use – because of selectivity reasons – MS or MS/MS detection. Stability of certain metabolites in plasma or urine could also be a big problem. Enzyme-induced splitting off phase-2 metabolites can lead sometimes to the parent drug. Also N-oxides are critical. Furthermore, with MS detection during ionization under higher temperatures some metabolites will be destroyed and mimic other metabolites or even the parent drug. Only the retention time on a RP system can give the necessary information for understanding. With N-oxides that sometimes have a similar retention time as the parent drug or a metabolite without N-oxide – this is especially dangerous.

Some examples for demonstrating the complexity:

Caffeine: If you want to determine Theophylline, Theobromine or Pentoxyphylline in plasma be very careful concerning consumption of Caffeine. For bioavailability studies stop the intake of Caffeine 1–2 days before starting your study. Caffeine will lead in the liver to about 20 similar Xanthenes even to Theophylline and Theobromine. Besides many food and stimulants consist of some unknown Caffeine!

Peppermint oil/Menthol: If you want to administer aetheric (essential) oils and want to determine Menthol in plasma this could become a nightmare. There is only some foodstuff that is free of Menthol. Also for tooth care or in body lotions

HPLC Methods for Clinical Pharmaceutical Analysis, First Edition. Hermann Mascher.
© 2012 Wiley-VCH Verlag GmbH & Co. KGaA. Published 2012 by Wiley-VCH Verlag GmbH & Co. KGaA.

often menthol is inside without any declaration (see our Menthol publication[13]). So volunteers for such a study need a specific meal without menthol for two days before the study starts and also the day after application - and no tooth cleaning or chewing gum at all!

Phytopharmaka: If you have to determine leading substances of usually orally applied phytopharmaka in plasma and urine you are in a terrible position. For example, plant substances with an aglycon with added sugars (so-called "glycosides") are often at least partially split off in the intestine and partially absorbed as aglyca.

There is a great quantity of natural competition from the usual nutrition. One example should show the practical problem: A 3-Glycoside of an active phytopharmakon is applied orally. But in the meal there is a lot of the same aglycon as 7-Diglykoside that is also taken with the lunch. In the intestine both glycosides were split off and the now "same" aglycon is absorbed. So, no bioequivalence comparison is possible.

Multimetabolites: For an old registered drug with 12 metabolites described we had the task to check, for a new submission to the authorities, the metabolism once more. After carefully checking samples from three animal models and humans (each plasma and urine) we found with our new highly sensitive and selective mass spectrometer (tandem MS/MS and Orbitrap XL) around 100 metabolites. Some of them are in the matrix in much higher concentration as the parent drug – so the determination of the parent drug or specific metabolites with unselective detectors like UV will lead to a disaster!

---

13 Drug Res. 51 (2001) 465-469 "Pharmacokinetics of Menthol and Carvone after Administration of an Enteric Coated Formulation Containing Peppermint Oil and Caraway Oil"

10
Internal Standards

In Section 5.1 you can find some remarks on this topic. The comparison between external and internal standardization shows advantages and disadvantages for both procedures. The decision as to which method you will use depends on the detector and the complexity of sample clean up. Optimal for MS or MS/MS is an isotopic labelled internal standard that has the same behavior as your substance but a sufficient mass difference. As a drawback it is not easy to get such a substance even for generic drugs. Besides they are rather expensive and they must be clean from the parent drug. The advantages are clearly seen: Such an internal standard has the same behavior during sample clean up and has the same retention time in HPLC as the parent drug. Also, ionization in MS should be very similar to the parent drug.

Similar substances as the parent drug but with similar lipophilic behavior or the same important functional groups can usually correct recovery during sample clean up, can correct dilution problems or even injections with varying volumes. But with MS detection through slightly different retention time or ionization behavior different matrix effects as with the parent molecule can occur[14][15].

By using UV or fluorescence detection the internal standard must be different from the parent drug with a different retention time with HPLC. But other behaviors should be similar – extraction or UV or fluorescence wavelength and intensity.

If these matrix effects are fairly similar with plasma of different subjects this should not be a big problem – but this is rarely the case. But another big problem can occur by using internal standardization: We observed with nonisotopic labelled internal standards that sometimes the results concerning precision, reproducibility or linearity were better only by using the peak area of the analyte (= external standardization) than the ratio analyte: internal standard (= internal standardization). This means that the internal standard can be unstable or have more matrix effects or have a nonreproducible recovery, for instance. The internal standard is also only a substance like the analyte. These effects can occur by using different

14 LC-GC Europe February 2002, 73-79 "Ion Suppression in LC-MS-MS-A Case Study"
15 Rapid Commun Mass Spectrom 17 (2003) 97-103 "Investigations of Matrix Effects in Bioanalytical HPLC MS assays: Application to drug discovery"

HPLC Methods for Clinical Pharmaceutical Analysis, First Edition. Hermann Mascher
© 2012 Wiley-VCH Verlag GmbH & Co. KGaA. Published 2012 by Wiley-VCH Verlag GmbH & Co. KGaA.

detectors. As a summary, different problems can come from internal standards and not from the analyte!

By using MS or MS/MS detection one should get usage, to have a look at the peak areas of the internal standards during a sequence or batch. So you can see if there is a difference between calibration, QC or real patient samples. But keep in mind if there is a complex sample preparation with different volume variations an internal standard is a "must" and might vary randomly just from sample preparation.

Important demands on internal standards:

- They must be stable in plasma, extraction solvents and buffers and in injection solutions for HPLC.
- They must have essential functional groups like the analytes to behave like the analytes through the sample preparation process.
- They must be similar lipophilically as the analytes to be extracted in the same way (LLE) and to have similar retention on RP-HPLC columns.
- They need similar pKa values so they do not get lost during different sample preparation steps.
- They need for MS and MS/MS detection – if isotopically labelled – enough mass difference from the analyte that there is no disturbing of the analyte trace.
- They must – when isotopically labelled – be distinctly labelled (e.g., 3 times deuterated) and are not allowed to have any traces of undeuterated substance, which would be the analyte.
- If they are middle polar the isotopic labelled part is not allowed to be too large, meaning the mass difference especially for deuterated internal standards should not be too large to the analyte. The reason is a difference in the lipophilic behavior. For example, a deuterated substance is more lipophilic than an undeuterated substance (= the analyte). Therefore, extraction and RP chromatography can be so different, that recovery during extraction is higher and retention is longer. Therefore, compensation of the matrix effect in MS can also be abolished.
- For unselective detection like UV, fluorescence, ELSD or ECD it is essential that for chromatographic reasons the internal standard has to have a different retention time from the analyte. Otherwise you cannot differentiate between analyte and internal standard. If you have similar retention times then the internal standard should elute after the analyte, otherwise small peaks of analyte may not be integrated in a correct manner after the (usually high – for safety reasons) internal standard. If you have a tailing internal standard peak this could be terrible.

11
Case Studies with Intensive Discussion for Each Substance

For practical reasons we have decided to cite the analytes in alphabetical order. To arrange in different groups like drugs, endogenous substances, phytopharmaka or vitamins didn't seem appropriate. For analytical reasons there is no difference between those analytes, coming from different origins. The important thing is the structure and each chemical/physical behavior of the substance. But there is one big difference in another way: drugs are not distributed ubiquitous, which is not true for substances from plants. Besides vitamins or other endogenous substances are always constituents of plasma or urine. This is a big challenge for validation and selectivity reasons. All the examples have a historical development in our lab. For all these analytes we developed methods during the last 30 years. These methods were for determination either from plasma/serum/urine or tissue. They were validated often under GLP guidelines but these guidelines changed over the years. We used those methods for preclinical and clinical samples. Having a look at those methods you can see the development of HPLC and all the HPLC detectors during these 30 years. We bought our first single quadrupole MS instrument 1995 just after introducing ESI and APCI ionization in the market. Thermospray – the ionization method before ESI and APCI could only be used in a small range for lipophilic substances like specific pesticides. ESI and APCI got a really broad application range for many substances in the pharmaceutical/clinical area. In 1998 we bought our first MS/MS tandem mass spectrometer after short and intensive testing. This opened for us a broad window – first for substances that have a poor UV behaviour therefore also a poor fluorescence, secondly for lower determination limits. HPLC-UV ends also for really UV-active substances with high ε values with wavelengths >300 nm with a determination limit of 1–5 ng/ml plasma – also with excellent sample clean up and enrichment. During the subsequent years we bought more and more tandem MS systems. At the moment we use two API4000 and two API5000. Furthermore, we have an ion trap – LTQ-Orbitrap-XL an FT instrument for metabolism, biomarker elucidation and also structure elucidation of unknown substances. This means before 1995 we had to think in another way from today (without having the chance of an MS instrument). Since 1998 we have had the opportunity of using all the options of a tandem MS/MS system for solving all of our bioanalytical questions.

HPLC Methods for Clinical Pharmaceutical Analysis, First Edition. Hermann Mascher
© 2012 Wiley-VCH Verlag GmbH & Co. KGaA. Published 2012 by Wiley-VCH Verlag GmbH & Co. KGaA.

Showing all those 20 examples in this chapter (Sections 11.1–11.20) we will not only show the solutions of these bioanalytical questions.

What really is important to me is guiding you through many thoughts concerning analytes, matrices, determination limits, solutions and pitfalls. At the end of this book around 100 substances are shown – each substance on one page. This is more a simple reference book. The composition on each page is showing rough determination limits for different detectors (UV, fluorescence, coulometric and MS/MS), MS/MS spectra, used sample preparations and citation of own published methods or from other publications. I tried to check all the abstracts of published methods and chose the realistic ones in my opinion – often with different detectors. So finally, you should be in the position to check your own detection options and make an estimation of the determination limits you can gain in your lab.

Let us start in alphabetic order.

11.1
Acetylcarnitine in Plasma

Acetylcarnitine belongs to the endogenous substances, more precisely to the acyl-carnitines.

MW: 203.2

C_{max}-value	protein binding (PB)	elimination half life $t_{1/2}$	pKa-value	Ionisation MS	Parent Ion
					Fragment Ion(s)
native value ca. 1000 ng/ml			-1.8	pos	204
					85

Discussion of Structure

Acetylcarnitine has no chromophores at all and therefore also no fluorescence. Using UV detection at 200 nm you might see a small peak for your pure substance for 50–100 ng (rough estimation). As there is no oxidizable functional group therefore, ECD detection is not possible. Without MS or MS/MS detection only derivatization of the –COOH is possible. Our experience showed us that derivatization of this molecule is really difficult. Despite our huge experience in the field of derivatization we could not confirm some publications of different ways of derivatization (in particular, Japanese groups did this) in our lab 15–20 years ago. There might be publications within the last 10 years that are appropriate to reproduce. The demand to analyze this substance came up in 2002 again (we worked in that area even before 1995). We worked on that topic with MS/MS detection and could solve it.

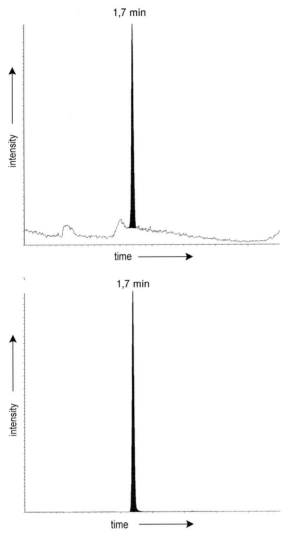

Chromatogram: 50 ng/ml in plasma, upper trace Acetylcarnitine (1.7'), lower trace internal standard (1.7')
Matrix: Plasma / serum

Sample Preparation
Acetylcarnitine is an amphoteric molecule and therefore not extractable with LLE. The short acylcarnitines have also almost no retention on RP-SPE. Weak cationic exchangers are an option using the quaternary nitrogen for interaction. If you have to determine all the acylcarnitines in a sample using a cationic exchanger would be a good chance to clean up the samples in a selective way.

The –COOH (carboxy-group) also offers a chance for sample clean up with a strong anionic exchanger. In our lab, different approaches with plasma and anionic exchanger for that carboxy group often resulted in low recovery with different analytes (the retention of the analyte was often relatively poor; the recovery was strongly depended on the matrix). So this way of sample preparation was not our preferred way.

Rough Method Description of Our Determination

For determination in plasma calibration samples in water were prepared. The calibration range was 50–10000 ng/ml. Each plasma sample contains acetylcarnitine therefore calibration samples were prepared in water. In addition, QC samples were prepared in water and in plasma, where also the native level was determined and subtracted for spiked concentration levels. With this concept you can have a fast impression of recovery behavior and matrix effects. 20 µL of calibration sample in water or plasma samples were mixed with 400 µL internal standard (D3-acetyl-carnitine in 90% acetonitrile). After vortexing (proteins were precipitated through acetonitrile) the samples were centrifuged. A small portion of the clear supernatant (from plasma) or clear solution (from water) was injected onto a bare-silica HPLC column with around 40% acetonitrile with TFA as mobile phase with MS/MS detection in positive mode.

Our publications concerning those topics (whole carnitine after hydrolyzes of the esters respectively, free carnitine) can be found in the footer[16][17][18].

11.2
Acetylcysteine in Plasma

Acetylcysteine belongs to endogenous substances with low native levels. It belongs to thiols and therefore it is a labile molecule especially in biological matrix.

MW: 163.2

16 Chemical Monthly 136 (2005), 1523-1533 "Carnitine in Pregnancy"

17 Chemical Monthly 136 (2005), 1425-1442 "Endurance exercise training and L-carnitine supplementation stimulates gene expression in the blood- and muscle cells in young athletes and middle aged subjects"

18 Clinical Chemistry 53 (2007), 717-722 "Long-term stability of amino acids and acylcarnitines in dried blood spots used for neonatal screening by tandem mass spectrometry"

solvent symbol	methanol ———	water — · — · —	0.1M HCl — — — —	0.1M NaOH · · · · · ·
absorption maximum				
$E_{1cm}^{1\%}$				
ε				

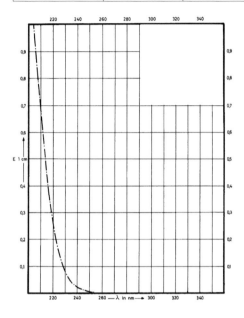

C_{max}-value	protein binding (PB)	elimination half life $t_{1/2}$	pKa-value	Ionisation MS	Parent Ion ——— Fragment Ion(s)
200 mg oral → 1.5 µg/ml	78 %	6 h	9.5	neg	162 ——— 84/33

Matrix: Plasma/serum

Discussion of Structure

Acetylcysteine has no chromophores at all and therefore also no fluorescence. Using UV detection at 200 nm you might see a small peak for your pure substance for 50–100 ng (rough estimation). Using ECD detection is really difficult for this molecule. Without MS or MS/MS detection only derivatization of the -SH or the -COOH is possible. Determination of the -COOH group would be possible by using precolumn derivatization. For GC, this derivatization is easy (methylation), for

HPLC and UV detection or fluorescence detection it is much more difficult, but still possible. We decided to use the -SH group for derivatization. We showed at a congress in Cordoba/Spain in 1989 a poster in three parts concerning "Postcolumn Derivatization of Thiols by O-Phthaldialdehyde":

Part 1: Common Investigations with Different Thiols;

Part 2: A New Very Sensitive and Selective HPLC-Determination of N-Acetylcysteine in Human Plasma;

Part 3: A New Highly Sensitive HPLC Determination of Captopril in Human Plasma.

A LLE as sample clean up is not possible with this very hydrophilic molecule.

Derivatization: Our idea was that using the NH_2 derivatization of, e.g., amino acids with OPA needs a thiol for ring closing resulting in a fluorophore. Why not use this reaction the other way round? Through systematic testing we checked different thiols used as drugs (among them also N-Acetylcysteine) in combination with different amino acids for that OPA reaction. The best reaction showed the highly hydrophilic glycine as primary amine. Glycine was simply given into the mobile phase (approx. 100 mg/L) and had no retention on the RP HPLC column. So it elutes in the same concentration from the HPLC column all the time. This eluate was mixed through a T-junction only with OPA in basic buffer (around pH 9). To give this postcolumn reaction enough time, a Teflon capillary of about 20 m was used after the T-junction before the fluorescence detector. The volume of this Teflon capillary was about 400 µL knitted, so that almost no peak broadening occurred.

Where do you see a peak in the chromatogram? Only when a thiol is eluting from the column. Then, this reaction takes place between thiol, glycine and OPA and forms a fluorescent molecule.

Useful tip 9: Thiols in plasma

Thiols will be easily oxidized in biological matrix. This process is often an -S–S- binding of two free -SH groups. In plasma there are many different thiols sometimes even in much higher concentrations:, e.g., Glutathione, Cysteine or peptides and proteins with free –SH groups. There are three ways in principle to solve these problems of oxidation:

You could restrict the oxidation by cooling the drawn whole blood and by fast centrifugation at 4 °C to get the plasma. The plasma must be frozen at –70 °C immediately. This process is a critical one for clinics to be standardized.

You could use antioxidants (which types are allowed in blood-drawing units?) just when blood drawing with the antioxidant in this vial (this means usually no vacuum vial is possible). Or it is possible immediately after blood drawing in the usual systems – by mixing afterwards with a solution of a specific antioxidant. For this process you have to open the drawing vial, to check the volume of the whole blood, to pipette a certain small volume of antioxidant in water and mix it immediately. To standardize this procedure is also hardly possible for routine clinical units.

Therefore, we decided to use the plasma that was frozen after the routine process in clinics.

Knowing from experiments in plasma N-Acetylcysteine has a short half-life in plasma. This means within minutes the levels of spiked N-Acetylcysteine was lowered significantly. But by using strong reducing agents putting into that plasma the complete amount of spiked N-Acetylcysteine could be found again. So this was a good way for the determination of whole N-Acetylcysteine in plasma.

Finally, only the selectivity concerning other thiols is important in the chromatogram.

Sample Preparation
Many details can be found in our publication[19].

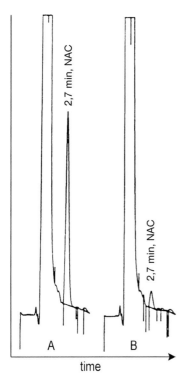

Chromatogram: A) 338 ng/ml / B) 25 ng/ml in plasma (2.7' each)

19 Biopharm. Drug Disposit. 12 (1991), 343-343 "Endogenous plasma N-Acetylcysteine and single dose oral bioavailability from two different formulations as determined by a new analytical method"

Rough Method Description of Our Determination

0.5 ml plasma was taken into a centrifuge tube after thawing and mixed with 30 µL 10% tributylphosphine (highly toxic) in methanol. After waiting for 30 min at 37 °C the proteins were precipitated with 0.15 ml 20% $HClO_4$. After centrifugation, a small part of the clear supernatant was chromatographed on C18 with a mobile phase of 4% ethanol/96% 0.1 M H_3PO_4 with 0.1% triethylamine (for peak shape) and 100 mg glycine per liter mobile phase. The online postcolumn derivatization was done with 41% 0.5 M NaOH/59% 0.1 M H_3BO_3 with 10 mg OPA per liter buffer. The fluorescence detector was used with 325 nm excitation and 475 nm emission.

11.3
Acyclovir in Plasma and Urine

Acyclovir is a synthetic purine (coming from Guanidine) that has antiviral activity.

MW: 225.2

solvent symbol	methanol ———	water — . — . —	0.1M HCl — — — —	0.1M NaOH · · · · · ·
absorption maximum	254 nm		255 nm	257 nm 264 nm
$E_{1cm}^{1\%}$	611		530	475 475
ε	13 800		11 900	10 700 10 700

C_{max}-value	protein binding (PB)	elimination half life $t_{1/2}$	pKa-value	Ionisation MS	Parent Ion ——— Fragment Ion(s)
200 mg oral → 400 ng/ml (usual 0.5 – 15 µg/ml)	9 – 33 %	1.5 – 6 h	-1.7	pos	226 ——— 152/135/110

Matrix: Plasma/serum and urine

Discussion of Structure

Acyclovir coming from Guanidine is similar to different endogenous molecules (e.g., Guanine, Guanosine, Hypoxanthine). Acyclovir is very hydrophilic – so LLE is impossible. Also, the retention on RP systems is weak. So, for sample clean up no selectivity against similar endogenous substances could be gained. We worked on this topic in 1989 – a long time before HPLC-MS instruments. Detection with ECD was not possible, so UV or fluorescence detection was the only way. If you take a look at the UV spectrum you see a good absorption behavior. But note one thing – the determination limit should be around 1 ng/ml plasma. So, UV detection was no option! What could give us a chance? Checking carefully the literature one small side remark in a scientifically weak publication became our chance. What was the remark? Acyclovir should have a weak fluorescence but fluorescence will be much stronger under more acidic conditions. Through many experiments we found a really strong fluorescence with $HClO_4$ in the mobile phase. $HClO_4$ also as an ion-pairing substance resulted in a good peak shape with enough retention for Acyclovir. Three flies at one blow!

This was the solution with enough selectivity concerning endogenous substances from plasma and urine.

Sample Preparation

Many details can be found in our publication[20]. Also, plasma levels can be found in another of our publications[21].

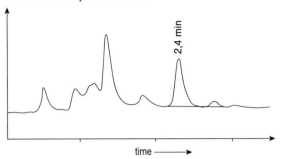

Chromatogram: 113 ng/ml in plasma, peak at 2.4'

Rough Method Description of Our Determination

1 ml plasma was mixed with 0.3 ml 3M $HClO_4$ for protein precipitation and followed by centrifugation. 20 µL of the clear supernatant was injected into the HPLC column. You may use only 0.1 ml plasma plus 30 µL $HClO_4$. We used a C18 column with 0.02M $HClO_4$ in water as mobile phase (with a step gradient with acetonitrile for column cleaning). The fluorescence detection used 260 nm excitation and 375 nm emission wavelength.

Urine samples were diluted with diluted $HClO_4$ and injected thereafter.

11.4
Caffeine in Plasma

Caffeine is a compound everybody knows and many people like it as a stimulant.

MW: 194.2

---

20 J. Chromatogr. 583 (1992), 122-127 "A new, high-sensitivity HPLC determination method for Acyclovir in human plasma, using fluorometric detection"

21 Drug Research 45 (1995), 508-515 "Pharmacokinetics and Bioavailability of different formulations of Acyclovir"

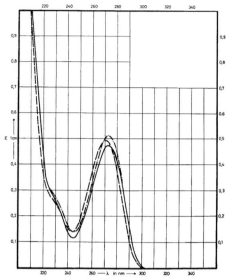

solvent symbol	methanol ———	water —·—·—	0.1M HCl —————	0.1M NaOH ······
absorption maximum	273 nm	273 nm	270 nm	273 nm
$E_{1cm}^{1\%}$	475	515	495	510
ε	9220	10 000	9610	9900

C_{max}-value	protein binding (PB)	elimination half life $t_{1/2}$	pKa-value	Ionisation MS	Parent Ion
					Fragment Ion(s)
100 mg oral → 3 µg/ml (usual 2 – 10 µg/ml)	35 %	2 – 10 h	14.0	pos	195
					138/110/83

Matrix: Plasma/serum

Discussion of Structure

You can see that this substance is highly hydrophilic, therefore, LLE is not the first choice. Although there are publications using extraction – even in our lab exists a determination for caffeine and metabolites of caffeine from urine – LLE has some drawbacks. The recovery is poor and only by using some tricky steps can the recovery be enhanced (usage of almost hydrophilic organic solvents, usage of the so-called "Aussalzeffekt"). Another drawback of this molecule is that there are many structural similar molecules existing in the body and therefore also in plasma or urine. It depends on the wanted determination limit – but a determination limit of 0.2 µg/ml can be gained with UV detection without many problems. If you only want to determine caffeine you may use SPE because of eliminating many disturbing substances – especially proteins and late eluting substances on RP. But SPE is not helpful for the selectivity of the given chromatogram.

Using MS or even MS/MS detection caffeine and its metabolites (around 20) can be determined relatively selective. The detection limit is around 2 pg on an API 4000 MS/MS system. But looking at selectivity even with MS/MS there are some problems especially with isomers of metabolites that have same fragments in common and have similar or the same retention times.

Sample Preparation
Using protein precipitation with acetonitrile from plasma you can get at least a clear supernatant for injection. But you have to evaporate the acetonitrile in a vacuum centrifuge for injection unto the HPLC column. Otherwise, only few microliters can be injected without strong peak broadening or bad peak shapes (both selectivity and sensitivity then suffering).

One important point must be mentioned: Be careful when looking for plasma or urine for spiking calibration and QC samples. Almost no one has no Caffeine intake. First, is Caffeine from coffee and tea consumption (decaffeinated coffee also contains Caffeine) and secondly: show me a person who does not have any kind of chocolate? An important constituent of chocolate is Theobromine that is metabolized partly to Caffeine.

Analyzing samples of human mother's milk we couldn't see any woman who has not a lot Caffeine, Theophylline or Theobromine in her milk!

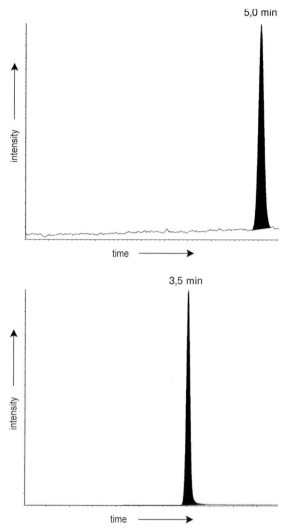

Chromatogram: 20 ng/ml in plasma, upper trace Caffeine (5.0'), lower trace internal standard (3.5')

11.5
Diazepam in Plasma

Diazepam is a substance that was the leader in the area of benzodiazepines. It still is applied nowadays also very often and has a huge misuse because of addiction.

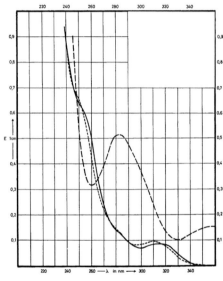

MW: 284.7

solvent symbol	methanol ———	water –·–·–	0.1M HCl – – – –	0.1M NaOH ······
absorption maximum	315 nm		360 nm 282 nm	310 nm
$E_{1cm}^{1\%}$	79		142 477	84
ε	2250		4040 13 580	2390

| C$_{max}$-value | protein binding (PB) | elimination half life t$_{1/2}$ | pKa-value | Ionisation MS | Parent Ion |
					Fragment Ion(s)
5 mg oral → 100 ng/ml (usual 200 – 500 ng/ml)	98 – 99 %	20 – 100 h	3.3	pos	285
					193/154/222

Matrix: Plasma/Serum

Discussion of Structure

Diazepam is a middle polar neutral molecule that can by extracted either by LLE or SPE from plasma. As the determination limit should be in the range of 10–20 ng/ml plasma HPLC-UV is barely possible. Neither ECD nor fluorescence detection can be used. Important phase-1 metabolites of Diazepam (Desmethyldiazepam, Oxazepam) are so different – namely more polar – that they can be separated easily from the parent molecule through RP chromatography – they elute earlier. Using MS or especially MS/MS all benzodiazepines can be determined with enough sensitivity.

Sample Preparation

By using only protein precipitation with acetonitrile and direct injection of the supernatant the determination with HPLC-UV is around 200–300 ng/ml plasma. This is because of sensitivity and especially selectivity reasons. HPLC-UV at 230 nm is rather unselective for biological samples. By using LLE or SPE sample clean up and enrichment can be made, therefore a LLOQ of 10–20 ng/ml with HPLC-UV can be reached. With HPLC-MS/MS you easily can get to 1 ng/ml plasma.

11.6
Diclofenac in Plasma

Diclofenac was one of the first substances used as NSAR (antirheumatic drugs). As this molecule has a specific pharmacokinetic behavior (very fast invasion – and also fast elimination half-life in humans) the blood-drawing time points are critical after oral application. If you allow only blood drawing each hour (which is usual after 1–2 h after oral ingestion of many drugs) it might happen that you only see the start of absorption, then 10–30% of maximum and the next time point will be still late in the elimination phase.

MW: 296.2

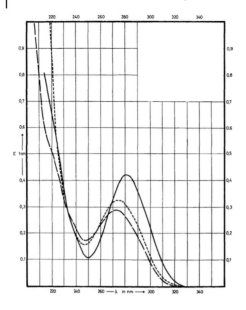

solvent symbol	methanol ———	water —·—·—	0.1M HCl — — — —	0.1M NaOH ······
absorption maximum	282 nm		274 nm	275 nm
$E_{1cm}^{1\%}$	425		288	327
ε	12 030		8150	9250

C_{max}-value	protein binding (PB)	elimination half life $t_{1/2}$	pKa-value	Ionisation MS	Parent Ion ——————— Fragment Ion(s)
50 mg oral → 800 ng/ml (usual 50 – 2500 ng/ml)	>99 %	1 – 2 h	3.1	pos	296 ——————— 214/250/178

Matrix: Plasma/serum

Discussion of Structure

Diclofenac possesses besides the carboxy functional group, also a slightly basic function. Extraction from diluted water solution under acidic conditions into middle polar organic solvents works well. But out of plasma under acidic conditions, it is variable. One of the first publications on this topic shows that clearly and helps to understand the problem of recovery[22]. Our first publication concerning Diclofenac was in 1989[23] with an HPLC-UV determination after extraction at 300 nm. This extraction in detail was under very strong acidic conditions and a re-extraction from organic solvent into a basic buffer[24]. The LLOQ was at 10 ng/ml plasma. At that time two different approaches for LLOQ were usual from biological fluids. The first approach was a signal: noise ratio of at least 3:1 or 5:1. At that time the printed chromatogram was taken and the noise measured with a lineal over 1-2 minutes (surely in a time range of no peak). The second approach was the extrapolated calibration curve and the point crossing the Y-axis (concentration axis). The determination level was at that point or higher.

Some years later we published methods with HPLC-MS/MS after extraction[25, 26] – surely much more sensitive. With MS/MS on an API 3000 the LOD is around 10 pg.

Sample Preparation
You might extract Diclofenac with LLE under strong acidic conditions (pH of acidified plasma around 1) or you use SPE where the pH is not really critical. Slightly acidic is better for complete retention but a natural pH of around 8 is not really bad. For more details see those publications.

Rough Method Description of Our Determination
1 ml plasma is mixed with 0.2 ml 4 M H_3PO_4 and 2 ml heptane/isoamylalcohol (97/3). After strong shaking and centrifugation the organic phase is drawn out and mixed with 0.02 M NaOH, and then centrifuged. The Diclofenac goes under acidic conditions undissociated into the organic phase and then as salt (after NaOH) into the water phase. 100 µL of that 0.02 M NaOH is injected into the HPLC-C18 system with 50% acetonitrile/50% 0.1 M H_3PO_4 as mobile phase with a UV detector at 300 nm.

Recovery for extraction and re-extraction was at a total of 87%, and the calibration range was 10–500 ng/ml plasma.

The following chromatogram shows the later developed HPLC-MS/MS determination only with extraction and evaporation of the organic phase.

22 Geiger et al.: J. Chromatogr. 111 (1975), 293-298 "Quantitative Assay of Diclofenac in Biological Material by GC"
23 Drug Design and Delivery 4 (1989), 303-311 "The pharmacokinetics of a new sustained-release form of Diclofenac sodium in humans
24 Clin. Pharmacol. Ther. 62 1997), 293-299 "Diclofenac concentrations in defined tissue layers after topical administration"
25 J. Pharm. Biomed. Anal. 33 (2003), 745-754 " Comparison of UV and tandem mass spectrometric detection for the high-performance liquid chromatographic determination of diclofenac in microdialysis samples"
26 Int. J. Clin. Pharm. Ther. 42 (2004), 353-359 "Topical skin penetration of diclofenac after single and multiple dose application of Voltaren Emulgel"

intensity: 26106 cps

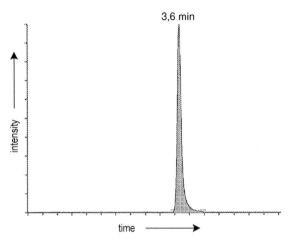

Chromatogram: 20 ng/ml in plasma (peak at 3.6')

11.7
Dihydralazine in Plasma

Dihydralazine is an old drug for blood pressure. We used it as an example of many problems in determination. The molecule is very reactive and therefore really unstable in biological fluids (building easily hydrazones with aldehydes and ketones). In our time where we often used derivatization (the time before MS/MS) we had used Dihydralazine as a reagent for postcolumn derivatization similar to what we did for vitamin B6 (published).

MW: 190.2

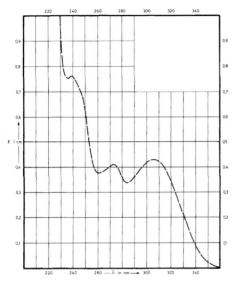

solvent symbol	methanol ———	water —·—·—	0.1M HCl —————	0.1M NaOH ······
absorption maximum			306 nm 274 nm 240 nm	
$E_{1cm}^{1\%}$ *			213 204 377	
ε			6130 5870 10 870	

* referred to dried substance

C_{max}-value	protein binding (PB)	elimination half life $t_{1/2}$	pKa-value	Ionisation MS	Parent Ion ——— Fragment Ion(s)
25 mg oral → 50 ng/ml		1 – 2 h	0.2	pos	191 ——— 129/102/174

Matrix: Plasma/serum

Discussion of Structure

Both hydralazine groups in the molecule look alarming for an expert in bioanalytical analysis. Besides both nitrogens in the ring (in the o-position to hydralazine) are not clearly defined concerning pKa. The UV spectrum looks fairly good (maximum at 306 nm). It would not be impossible that fluorescence could occur under specific conditions. The alarming aspect is the reactivity of both hydralazine groups. All published papers dealing with the bioanalytical determination of Dihydralazine derive from the years around 1980. All they determine is the total hydrazone after some reactions. First, all the reaction products in plasma (Dihydralazine disappears very fast from plasma) are rebuilt with specific reagents and then a derivatization step was done. With much effort we tried to understand these published reactions – using HPLC-MS/MS. We wanted to understand the reactions to optimize the procedures. Finally, we were disappointed. It was not possible to get Dihydralazine back after destruction in plasma (only 5–10%, contrary to remarks (e.g., complete) in those publications). So, finally the only option was to slow down the degradation in plasma after blood drawing. The half-life in plasma at room temperature was around 10 min. The procedures of stabilization are described in our publication[27]. But this was only the first step to overcome the problems for the determination. Gaining from plasma and stabilizing in the injection solution were the next critical steps. Also, it was necessary to find a fitting internal standard. But this was not the end of our problems: During HPLC separation we observed a bad carryover effect and a bad peak shape. These two problems were finally solved by putting a similar substance into the mobile phase (Isonicotine Hydralazine 20 mg/L mobile phase). Also, the ionization process in MS/MS showed tailing peaks and no linearity – both solved through Isonicotine Hydralazine in mobile phase too.

Rough Method Description of Our Determination

Because of stability reasons only 6 samples each were thawed together and prepared, (within 5 min thawing, pipetting 200 µL into a prepared vial). This vial contains 50 µL of internal standard and 50 µL of 20% TCA for protein precipitation. 20 µL clear supernatant was injected (autosampler at 5 °C).

For plasma samples gained in a clinic the following procedure is necessary: To the whole blood (plasma gained in a vial within 5 min) 1% (v/v) of 5% DTT solution (Dithiothreitol) in water has to be added, mixed and immediately 10 min centrifugation at 4 °C must be applied, and then skin plasma immediately and shock freeze it at –70 °C.

HPLC

The separation is on a C18 column with a gradient from 20 mM TCA (trichloroacetic acid) in water to 20 mM TCA in acetonitrile both with 20 mg Isonicotine

27 J. Pharm. Biomed. Anal. 43 (2007), 631-645 "Method Development of Dihydralazine with HPLC-MS/MS – an old but tricky substance – in Human Plasma

Hydralazine/L. The detection method used was MS/MS on an API 3000 in positive mode.

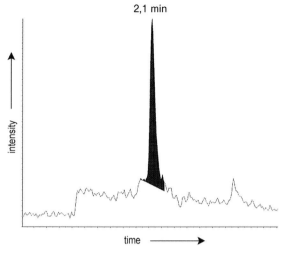

2,1 min

intensity →

time →

Chromatogram: 0.5 ng/ml in plasma (2.1 min).

11.8
Duramycin (Moli1901) in Plasma

Duramycin is a peptide consisting of 19 amino acids. It is used in the therapy of cystic fibrosis in an aerosol for inhalation.

MW: 2013

C_{max}-value	protein bin-ding (PB)	elimination half life $t_{1/2}$	pKa-value	Ionisation MS	Parent Ion
					Fragment Ion(s)
>1 – 10 ng/ml				pos (2⁺⁺)	1007
					1007

Matrix: Plasma/serum

Useful tip 10: Analysis of peptides

Peptides as pure substances can be analyzed in different ways by HPLC and using different detection modes. Usually, on RP a gradient from TFA in water to TFA in acetonitrile is used. The UV detection of peptides without tyrosine and Tryptophan will be done at 210–220 nm. If there are both or one of these aromatic amino acids in a peptide the wavelength can be changed to 270–285 nm (Tyrosine 270 nm, Tryptophan 285 nm). This definitely enhances the selectivity. At 210–220 nm each gradient will give a rising base line. But the sensitivity at 270–285 nm will be lower as only some per cent of the amino acid contest are the two aromatic amino acids. Tryptophan has weak fluorescence under specific conditions (pH), which is helpful for selectivity. As the excitation and emission wavelengths are close together (resulting in a noisy baseline) no real sensitive determination from plasma can be gained.

Phenylalanine (also an aromatic amino acid) has a maximum at 260 nm with a fairly low ε value. Therefore, 210–220 nm is the better choice.

With MS or MS/MS as detector ESI is the ionization source of choice. But be careful, depending on the size of the molecule multiply charged ions can occur. MS alone (single quadrupole) is usually not sensitive and selective enough for determination from plasma.

By using MS/MS you can choose two approaches:

- By stressing the selectivity, MRM (multiple reaction monitoring) is the best way. Some peptides have low abundance in fragment ions and they are sometimes not really selective, as many peptides consist of only a few different amino acids.
- By stressing the sensitivity sometimes a so-called "pseudo-MRM" is the best choice. "Pseudo-MRM" means both quadrupoles are fixed on the parent ion (no matter if singly or multiple charged) and the tension of fragmentation is so low that no fragmentation of the chosen peptide occurs. As some peptides need high energy for fragmentation with lower collision energy (respectively, a voltage applied in the fragmentation region of the mass spectrometer) many coeluting peaks may be destroyed or the chemical background will be lowered. Therefore, some noise or disturbing signals are diminished at the second quadrupole. Selectivity is lower with "pseudo-MRM", the baseline is often much higher than with MRM, but sensitivity (as signal-to-noise ratio) can be better by a factor of 5–20 with fitting peptides.

We got a LOD for Duramycin with "pseudo-MRM" of 25 pg.

Discussion of Structure

These 19 amino acids with some additional (= free) acidic and basic groups are generally fitting for electrospray ionization (ESI).

Sample Preparation

Usually, we precipitate proteins with acetonitrile when we determine peptides (up to 20–30 kDa) from plasma. This precipitation is the first cleaning step. For extremely lipophilic peptides we use other procedures. For Duramycin the recovery after protein precipitation with acetonitrile was around 10–20%. After carefully checking different precipitation methods we were sure that this is not a question of solubility (which can occur with more lipophilic peptides). Also some larger hydrophilic peptides might have solubility problems in 50% or 67% acetonitrile (which is the clear supernatant after protein precipitation).

The reason for low recovery of Duramycin was a kind of "protein binding". Through many experiments we discovered that the combination of urea and ammonia in high concentrations lead to a high recovery. Fortunately, Duramycin can stand this brachial treatment. For further details see in our publication[28].

So finally, we got a determination limit of 1 ng/ml plasma. Also different HPLC columns and mobile phases had to be checked carefully to get a "clean window" in the chromatogram for the Duramycin peak.

Rough Method Description of Our Determination

100 μL plasma is mixed with 80 μL 10 M urea solution and 20 μL 25% ammonia and has to wait for 15 min. Afterwards, 500 μL acetonitrile (including internal standard) was used for protein precipitation. The clear supernatant was chromatographed on a C18 column and detected with MS/MS with ESI positive ionization.

---

28 Naunym Schmiedebergs Arch Pharmacol. 378 (2008), 323-333, "Pulmonary pharmacokinetics and safety of nebulized duramycin in healthy male volunteers"

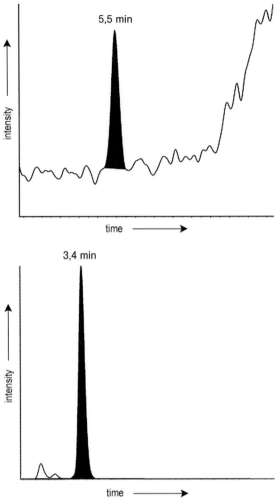

Chromatogram: 1 ng/ml in plasma, upper trace Duramycin (5.5'), lower trace internal standard (3.4')

11.9
Fluticasone Propionate in Plasma

Fluticasone Propionate is used as an anti-inflammatory drug in asthma therapy. As the molecule is applied with a spray in low amounts of up to 500 µg the systemic concentration in plasma is really low. Therefore, an LLOQ of better than 20 pg/ml is necessary.

MW: 500.6

solvent symbol	methanol ———	water — · — · —	0.1M HCl — — — — —	0.1M NaOH · · · · · ·
absorption maximum	237 nm			
$E_{1cm}^{1\%}$	372			
ε	18 600			

C_{max}-value	protein binding (PB)	elimination half life $t_{1/2}$	pKa-value	Ionisation MS	Parent Ion — Fragment Ion(s)
>50 – 100 pg/ml after inhalation	81 – 95 %	3.5 h		pos	501 — 313/293/205

Matrix: Plasma/serum

Discussion of Structure

Facing the necessity of such low determination limits (<20 pg/ml plasma) only MS/MS is the detection method of choice. The molecule is highly lipophilic with no acidic or basic functional group. Therefore, LLE is the method of choice. A second extraction to eliminate lipophilic acids or bases can be applied (see also Sections 1.1 and 3.3.1) to eliminate possible disturbing substances.

Sample Preparation

We used the following extraction procedure[29].

Rough Method Description of Our Determination

0.5 ml plasma will be mixed with internal standard and 2 ml diisopropylether. After shaking and centrifugation the sample will be frozen at –70 °C for about 10 min. Then, the organic phase will be decanted (remained liquid) and after adding 50 µL DMSO the organic phase will be evaporated at 50 °C. The residue including DMSO will be mixed with 65% acetone, mixed and injected onto the C18 HPLC column.

Detection was done with MS/MS and APPI ionization in negative mode. The calibration range was 3–1000 pg/ml or 10–1000 pg/ml.

HPLC-MS/MS

From our experience for such a low determination you need a really sensitive tandem MS system. The best systems 10 years ago were not able to get such low LLOQs. Fluticasone Propionate is not really well ionized with ESI but much better with APCI (LOD ca. 2 pg on API 4000) or even APPI (factor 3–5 more sensitive when mixing the dopant into the mobile phase). For more details see our publication.

29 J. Chromatogr. B 869 (2008), 84-92 "Sensitive simultaneous determination of Ciclesonide, Ciclesonide-M1-metabolite and Fluticasone Propionate in Human Serum by HPLC-MS/MS with APPI ionisation"

Useful tip 11: APPI (atmospheric-pressure photoionization) for MS
Through intensive testing of APPI interfaces we recognized that very lipophilic or sulfur containing molecules are really good fitting to APPI. It was our idea (we were the first ones who published this) not to pump the dopant into the interface through an extra pump (the producer recommends this) but mix it into the mobile phase. As acetone is similar to toluene as dopant acetone has similar reversed-phase behavior to acetonitrile. So, we use about 10–20% acetone instead of aceto-nitrile and get around a factor of 4–5 smoother silent baseline. This means a factor of 4–5 better signal-to-noise ratio[30].

Finally, we used APPI negative (LOD ca. 1 pg on an API 3000).

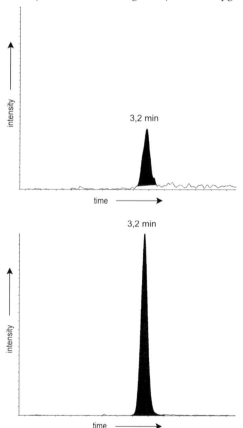

Chromatogram: 3 pg/ml in plasma, upper trace Fluticasone Propionate (3.2'), lower trace internal standard (3.2')

30 J Chromatogr B 869 (2008), 84-92 „Sensitive simultaneous determination of Ciclesonide, Ciclesonide-M1-metabolite and Fluticasone Propionate in Human Serum by HPLC-MS/MS with APPI ionisation"

11.10
Hydroxytriamterene Sulfate and Triamterene in Plasma and Urine

Hydroxytriamterene Sulfate is an important metabolite of Triamterene as a potassium-retaining drug. We published the described method in 1994[31].

MW: 253.3

solvent symbol	methanol ———	water — ·— ·—	0.1M HCl — — — —	0.1M NaOH ······
absorption maximum	266 nm		357 nm 250 nm	270 nm
$E_{1cm}^{1\%}$	568		855 634	559
ε	14 390		21 660 16 060	14 160

31 J. Liqu. Chromatogr. 17 (1994), 1577-1585 "Simple and Fast HPLC-Method for the Determination of Triamterene and Hydroxy-triamterensulphate in Plasma and Urine"

C_{max}-value	protein binding (PB)	elimination half life $t_{1/2}$	pKa-value	Ionisation MS	Parent Ion
					Fragment Ion(s)
50 mg oral → 300 ng/ml (usual 50 – 160 ng/ml)	60%	2.5 – 10 h	7.0/9.2	neg	296
					205/269/78

Matrix: Plasma/serum and urine

Discussion of Structure

It is easy to see that Triamterene is a very hydrophilic molecule (Hydroxytriamterene sulfate is even more hydrophilic). Therefore, LLE can not be used for sample clean up. Both substances are so hydrophilic that they have almost no retention on RP columns. Using an amino-column for HPLC with $HClO_4$/Triethylamine in water as mobile phase both substances can clearly be separated within 2 min.

UV detection at 360 nm could be used. But for more sensitivity we tend to use

---

Useful tip 12: Testing of fluorescence detectors

We knew Triamterene very well for testing different fluorescence detectors. Before MS/MS we always looked for the most sensitive UV and fluorescence detectors on the market. Most sensitive always means signal-to-noise ratio to us. So we tested all new fluorescence detectors coming to the market. We used 2 substances for this: Triamterene for absolute sensitivity and Propranolol because of much lower excitation wavelength. For Triamterene we got an LOD of a few femtograms for the most sensitive detectors. Even nowadays we would use for the determination of Triamterene from plasma fluorescence detection instead of our tandem-MS systems.

---

the most sensitive and selective detection method, therefore we used fluorescence.

Rough Method Description of Our Method

0.2 ml plasma was mixed with 0.6 ml water and then injected. With urine we used 20 µL diluted with 2 ml of water. Chromatography was done on Spherisorb amino with 0.01 M HClO4/0.002 M triethylamine and 0.1 M ammonium acetate in water as mobile phase. Detection was fluorescence with 360-nm excitation and 436-nm emission wavelengths. No internal standard was used because of simple sample handling and stable chromatography and detection.

Chromatogram: 19 ng/ml Triamterene (0.5') / 169 ng/ml Hydroxytriamterene sulfate (1.3') in plasma each

time ⟶

11.11
Ibuprofen in Plasma (also Enantiomeric Separation)

Ibuprofen – a NSAR is usually administered as racemate. Partly, it is as the S-enantiomer under the brand Seractil^R also on the market. Altogether, it makes sense to do an enantioselective determination in plasma.

MW: 206.3

solvent symbol	methanol ———	water – · – · –	0.1M HCl – – – – –	0.1M NaOH · · · · · ·
absorption maximum	272 nm 264 nm 258 nm			272 nm 264 nm 258 nm
$E^{1\%}_{1cm}$	11.2 14.5 11.3			15.4 18.4 15.0
ε	230 300 233			320 380 310

C$_{max}$-value	protein binding (PB)	elimination half life t$_{1/2}$	pKa-value	Ionisation MS	Parent Ion
					Fragment Ion(s)
200 mg oral → 20 µg/ml (usual 15 – 30 µg/ml)	99 %	2 h	3.8	neg	206
					161/91/119

Matrix: Plasma/serum

Discussion of Structure

In Section 1.1 there is an intensive discussion about the structure and the options for determination.

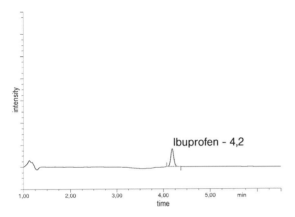

Chromatogram: 3.2 µg/ml in plasma (4.2'), HPLC-UV 214 nm

From our publications you may take many details concerning enantioselective determination in plasma[32][33].

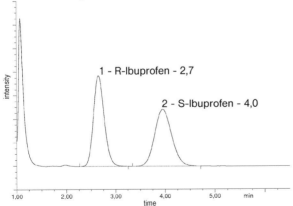

Chromatogram: 1.3 µg/ml per enantiomer (R 2.7', S 4.0') in plasma, HPLC-Fluorescence (225 nm / 290 nm)

32 Eur. J. Clin. Pharmacol. 48 (1995), 505-511 "Comparison of the bioavailability of dexibuprofen administered alone or as part of racemic ibuprofen"

33 Drug Research 47 (1997), 750-754 "Preliminary Toxicokinetic Study with Different Crystal Form of S (+)-Ibuprofen (Dexibuprofen) and R,S-Ibuprofen in Rats"

11.12
Minocycline in Plasma

Minocycline belongs to the Tetracyclines that could also be helpful concerning dermal problems.

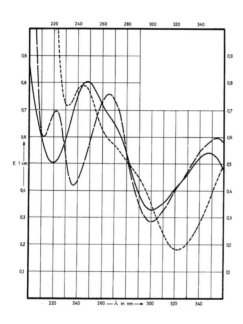

MW: 457.5

solvent symbol	methanol ———	water — · — · —	0.1M HCl – – – –	0.1M NaOH · · · · · ·
absorption maximum	348 nm 248 nm		354 nm 265 nm	385 nm 244 nm
$E_{1cm}^{1\%}$	256 385		290 370	335 383
ε	12 630 19 000		14 300 18 280	16 560 18 900

C_{max}-value	protein binding (PB)	elimination half life $t_{1/2}$	pKa-value	Ionisation MS	Parent Ion
					Fragment Ion(s)
50 mg oral → 1.5 µg/ml	75 %	11 – 26 h	2.8/5.0/ 7.8/9.5	pos	458
					441/352/283

Matrix: Plasma/serum

Discussion of Structure

Minocycline like all Tetracyclines has several pKa values.

This produces problems in sample clean up and also in chromatography. Minocycline is one of the most lipophilic Tetracyclines and therefore elution on RP systems is later than Tetracycline, Chlortetracycline or Doxycycline. At the time we had to deal with this molecule (1991) RP materials were not as well deactivated than nowadays. Besides the stability of the RP materials was poor when being outside the suggested pH range of 3–8. The originally poor peak shape could be strongly improved by using perchloric acid and triethylamine as constituents in the mobile phase. LLE is with all Tetracyclines hardly possible whereby Minocycline as a more lipophilic Tetracycline forms an exception. The detection was UV at 350 nm that eliminates many endogenous substances that have UV absorption at lower wavelengths only. ECD was no option (although the analyte possesses an oxidizable phenolic group) because of the complex mobile phase that could not be used with ECD. MS or MS/MS are nowadays in combination with good endcapped RP materials, a very good option.

Sample Preparation

As Tetracyclines usually have 4 pKa values you could hardly find a pH range where the molecules are fairly undissociated. Only under these circumstances you might think of LLE. SPE can be a good option to eliminate proteins and many other substances and it is also helpful for sample enrichment. In our published paper you can find a simple sample preparation[34].

Rough Method Description of Our Determination

KH_2PO_4 solution is added to plasma to set the pH at about 5. Then, LLE with ethyl acetate is done. The organic phase is re-extracted with a small volume of 0.02 M HCl. Minocycline goes as salt into the water phase during re-extraction. Then, a part of this phase is injected into the HPLC.

34 J. Chromatogr. A 812 (1998), 339-342 "Determination of minocycline in human plasma by HPLC with UV detection after liquid-liquid extraction

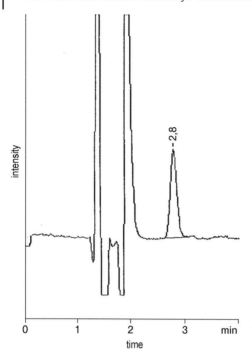

-2,8

intensity

0   1   2   3   min

time

Chromatogram: 148 ng/ml in plasma (2.8')

11.13
Norfloxacine in Plasma and Urine

Norfloxacine is a derivative of Chinolon and has been on the market for many years as a potent antibiotic. As it is an amphoteric molecule almost no freedom for sample clean up is given.

HN

CH₃

N

N

F

COOH

O

MW: 319.3

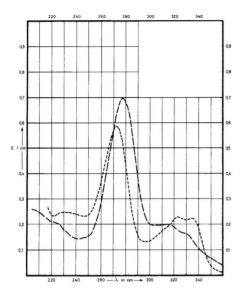

solvent symbol	methanol ———	water — . — . —	0.1M HCl – – – – –	0.1M NaOH . . . . . .
absorption maximum			315 nm 277 nm	335 nm 323 nm 272 nm
$E_{1cm}^{1\%}$			383 1350	430 440 1130
ε			12 200 43 100	13 700 14 000 36 100

C_{max}-value	protein binding (PB)	elimination half life $t_{1/2}$	pKa-value	Ionisation MS	Parent Ion ——— Fragment Ion(s)
400 mg oral → 1.3 µg/ml	14 %	3 – 4 h		pos	320 ——— 302/233/276

Matrix: Plasma/serum and urine

Discussion of Structure

As a Piperazin-base and with a carboxy functional group the molecule is relatively hydrophilic and is also amphoteric. During our practical work (1991) with this molecule we recognized stability problems and during chromatography problems with peak shape and carryover. Finally, usage of acetonitrile for protein precipitation and much $HClO_4$ in the mobile phase was helpful. Looking at the UV spectrum 280 nm is a good wavelength for getting a determination limit of 10–20 ng/ml plasma. If you need lower determination limits under certain circumstances (strongly pH-dependent) fluorescence can be successfully used. Nowadays, MS and MS/MS are a really good choice by looking at the structure of the molecule.

Sample Preparation
In our publication[35] you will find many details for determination in plasma and urine.

Rough Method Description of Our Determination
0.5 ml plasma was mixed with 0.5 ml acetonitrile, vortexed and centrifuged. The urine was simply diluted. HPLC separation was done on C18 with 30% methanol and 70% 0.1 M $HClO_4$/0.02 M triethylamine with fluorescence detection (excitation 300 nm, emission 450 nm).

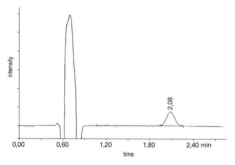

Chromatogram: 91 ng/ml in plasma (2.1')

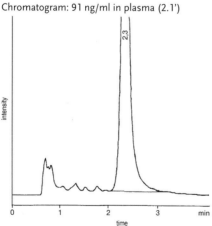

Chromatogram: 410 µg/ml in urine (2.3')

35 J. Chromatogr. A 812 (1998), 381-385 "Determination of norfloxacin in human plasma after protein precipitation by HPLC and fluorescence detection"

11.14
Paclitaxel in Plasma, Urine and Tissue

Paclitaxel is an ingredient of yew and is used in cancer therapy.

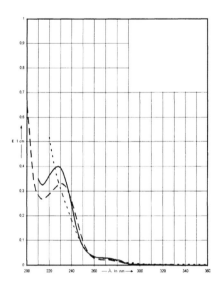

MW: 853.9

solvent symbol	methanol ———	water – · – · –	0.1M HCl – – – – –	0.1M NaOH · · · · · ·
absorption maximum	228 nm		231 nm	
$E_{1cm}^{1\%}$	403		333	
ε	34 400		28 400	

C_{max}-value	protein binding (PB)	elimination half life $t_{1/2}$	pKa-value	Ionisation MS	Parent Ion
					Fragment Ion(s)
50 – 500 ng/ml	95 – 98 %	3 – 50 h	3.3	pos	854
					286

Matrix: Plasma, urine and tissue

Discussion of Structure

Paclitaxel has a rather complicated structure with hydrophilic and hydrophobic parts. Therefore, be careful about solubility in hydrophilic and hydrophobic solvents. At the time point we were confronted with this topic (2006) a rather low determination limit especially in tissue was necessary. Therefore, only MS/MS came into question. During our first HPLC-MS/MS tests we recognized that this molecule needs a special treatment. First, we gained only a LOD of about 20 pg. Further systematic checks with different mobile phases (solvents, salts and additives) and different ionization sources (ESI, APCI and APPI, positive and negative) led to a LOD of 0.7 pg with ESI positive.

Be careful with the ionization temperature – it is really important.

Rough Method Description of Our Determination

To 50 µL plasma the internal standard (Docetaxel) was added and 1 ml diisopropylether. After shaking and centrifugation the plasma phase was frozen out at –70 °C for 10 min. Then, the organic phase was decanted and evaporated. After redissolving, a portion was injected into HPLC on C18 with MS/MS detection (ESI in positive mode).

For urine, the same procedure was used.

For tissue analysis 5 mg of the homogenized tissue was used.

As we had recovery problems for specific tissue types we used some Triton X100 as detergent for all types. To about 5 mg homogenized tissue 100 µL 50% ethanol, 20 µL 5% Triton-X100 in ethanol, 50 µL internal standard in ethanol and 300 µL water were mixed and extracted with 4 ml diisopropylether. Further procedure was as with plasma.

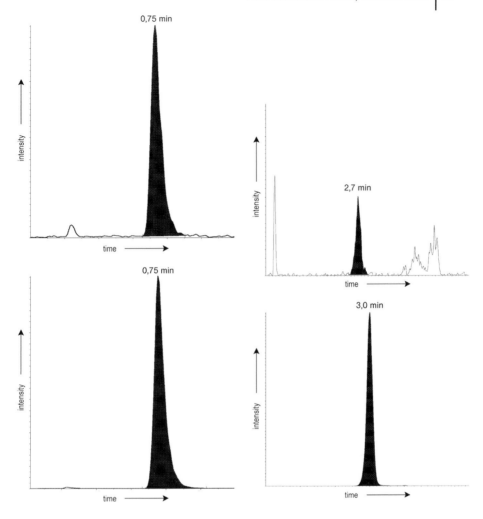

Chromatogram: 10 ng/ml in plasma, upper trace Paclitaxel (0.75'), lower trace internal standard (0.75')

Chromatogram: 10 ng/g in aorta tissue, upper trace Paclitaxel (2.7'), lower trace internal standard (3.0')

11.15
Paracetamol (Acetaminophen) in Plasma

Paracetamol is a classical drug for pain and fever.

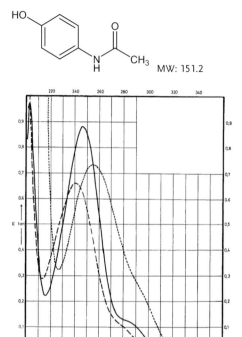

MW: 151.2

solvent symbol	methanol ———	water —·—·—	0.1M HCl —————	0.1M NaOH ······
absorption maximum	247 nm		240 nm	255 nm
$E_{1cm}^{1\%}$	850		642	710
ε	12 850		9710	10 740

C_{max}-value	protein binding (PB)	elimination half life $t_{1/2}$	pKa-value	Ionisation MS	Parent Ion ——— Fragment Ion(s)
10 – 20 µg/ml (usual therapeutic range toxic/poisonous >300 µg/ml) 500 mg oral → 5 µg/ml	<5 %	1.5 – 3 h	9.5	pos	152 ——— 110/65/93

Matrix: Plasma/serum

Discussion of Structure

Paracetamol is a phenolic compound with a short hydrophilic side chain. Therefore, sample preparation and clean up does not have many options. At the time we were confronted with this topic (1995) we only had UV, fluorescence or ECD as detectors.

Looking at the UV spectrum and the strong ε value at 247 nm and the need for only 50 ng/ml plasma it was evident to use UV as detector. Chromatography on RP was rather simple. During method development we observed high broad peaks after about 10 injections (late elution of endogenous substances). For such problems a step gradient could be an optimal solution but at that time not all of our HPLC systems had that option.

Furthermore, we were injecting in some systems by hand (not using an autosampler) and we used an isocratic run because this is much faster than a gradient. As phenols can be oxidized easily with a coulometric detector (ECD) we used that detector and had no selectivity problems at all. From our experience MS or MS/MS could be used for such a high LLOQ without any problem. If you have to go down with that level Paracetamol is not a really sensitive substance and with that low molecular weight of 151 there might be many problems with MS because of selectivity reasons.

Sample Preparation

Our publication – not in a peer-reviewed journal – gives only a few of the method details[36].

Rough Method Description of Our Determination

To 0.5 ml plasma 0.5 ml acetonitrile is added for protein precipitation. After vortexing and centrifugation part of the clear supernatant was mixed with diluted $HClO_4$ and 20 µL were injected onto the HPLC C18 column. The mobile phase consisted of 20% methanol/80% 0.02 M $HClO_4$ in water. Detection was done by ECD at +0.25 V (ESA, coulometric detector). The determination limit was 50 ng/ml plasma.

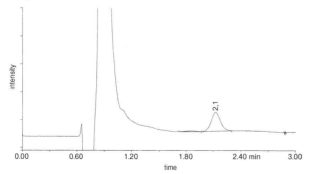

Chromatogram: 0.54 µg/ml in plasma (2.1') (electrochemical detection)

---

36 Forum DR.MED 17 (1996), 53-58 "Optimierte Bioverfügbarkeit von Paracetamol am Beispiel von ViCetamol-Brausepulver"

11.16
Pimelic Acid in Plasma and Urine

Pimelic acid is a short dicarbonic acid that is an endogenous substance in animal and human organisms,, e.g., on the skin, which gave many problems during method development. At the beginning we wondered about a high fluctuation of endogenous levels, although they were self-made by the sweat of the skin of our hands, during signing the tubes or drawing off or on of a cap.

HO ... OH   MW: 160.2

C$_{max}$-value	protein binding (PB)	elimination half life t$_{1/2}$	pKa-value	Ionisation MS	Parent Ion
					Fragment Ion(s)
usual 2 – 10 ng/ml				neg	159
					97

Matrix: Plasma/serum and urine

Discussion of Structure
As a relatively short dicarbonic acid Pimelic acid takes an unpleasant intermediate position in the lipophilic behavior. One could state "neither fish nor meat". Therefore, the extraction gets difficult, because during method development pure solutions in water were not comparable to urine and not to plasma at all. Even the usage of a deuterated internal standard (work done during 2007) for MS/MS detection gave no clear results concerning compensation. This internal standard was 4-fold deuterated and therefore distinctly more lipophilic than the undeuterated Pimelic acid. This was a problem concerning recovery during LLE and also with HPLC-MS/MS because of not having the same retention time. As we needed a low LLOQ it took us much effort to also get a low LOD. During the HPLC-MS/MS development we recognized (after days of work) that enhancement of ionization temperature resulted in much better signal-to-noise levels but unfortunately the internal standard lost during ionization part of the deuteration, which then pretended to be the endogenous Pimelic acid. To recognize such an effect is quite difficult. As a recapitulation we saw under the worst conditions (ionization temperature, composition of mobile phase, pollution of the spray) that approx. 5% of the deuterated internal standard was destroyed to Pimelic acid with APCI.

Rough Method Description of Our Determination
To 50 μL plasma 150 μL internal standard solution in methanol was added for protein precipitation. 15 μL of the clear supernatant were injected into HPLC-MS/MS with (finally) ESI negative. HPLC separation was done on an amino-phase with 60% methanol/40% 120mM formic acid.
Calibration range for plasma 0.5–200 ng/ml.

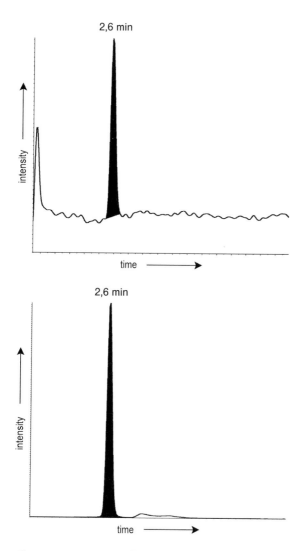

Chromatogram: 0.95 ng/ml in plasma, upper trace pimelic acid (2.6'), lower trace internal standard (2.6')

The urine samples were diluted with internal standard in methanol. The calibration range for urine was 20–20.000 ng/ml.

11.17
8-Prenylnaringenin in Plasma and Different Types of Tissues

8-Prenylnaringenin is a natural compound existing in hop plants. It is one of the strongest phyto-estrogens.

MW: 340.0

C_{max}-value	protein binding (PB)	elimination half life $t_{1/2}$	pKa-value	Ionisation MS	Parent Ion
					Fragment Ion(s)
				pos	341
					165/285

Matrix: Plasma/serum and different types of tissues

Discussion of Structure

We were confronted with this topic in 2002 and needed to gain determination limits as low as possible. Therefore, MS/MS was the detection of choice.

I want to stress again that the demand for lower determination limits is not only a question of sensitivity of the detector. It is a question of the whole method (sample preparation, HPLC separation, sensitive and selective detection). Everybody who is not such an expert in that bioanalytical area could easily forget this aspect. Lowering the determination limit by a factor of 10 enhances the number of endogenous "enemies" dramatically. Therefore, the selectivity must also be much better by working at such low levels. In addition, the danger of matrix effects with MS/MS increases considerably.

8-Prenylnaringenin and similar substances (e.g., 6-PN, Xanthohumol, Isoxanthohumol) can be easily ionized. The LOD is around 10 pg.

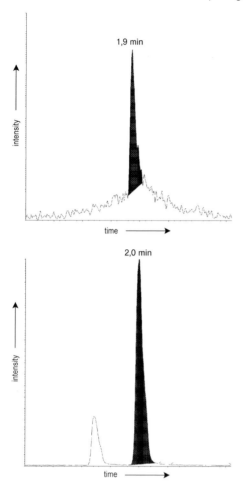

Chromatogram: 0.5 ng/ml in plasma, upper trace 8-Prenylnaringenin (1.9'), lower trace internal standard (2.0')

Rough Method Description of Our Determination

0.3 ml plasma, 0.5 ml 0.9% NaCl solution and internal standard were mixed with 4 ml diethyl ether and shaken. After centrifugation, freezing out the plasma phase, and decanting, the organic solvent was evaporated. After redissolving, the sample was injected into HPLC-C18 column with MS/MS (APCI positive mode) detection. The calibration range was 0.5–6.000 ng/ml plasma.

Tissue (rat liver/rat brain): After homogenization (Dismembrator in liquid nitrogen) an approx. 10% suspension in water was made and then extracted according to plasma.

11.18
Silibinin in Plasma

Silibinin is a natural substance coming from milk thistle. Pharmaceutical preparations contain diastereomers of Silibinin and Isosilibinin.

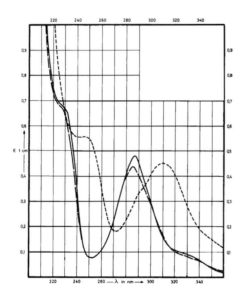

MW: 482.4

solvent symbol	methanol ———	water – · – · –	0.1M HCl – – – –	0.1M NaOH · · · · · ·
absorption maximum	287 nm		286 nm	310 nm
$E_{1cm}^{1\%}$	450		411	423
ε	21 700		19 800	20 400

C$_{max}$-value	protein binding (PB)	elimination half life t$_{1/2}$	pKa-value	Ionisation MS	Parent Ion
					Fragment Ion(s)
100 mg oral → 110 ng/ml free / total 2500 ng/ml (Glucuronid)		2.5 h	1.9	neg	481
					301/125

Matrix: Plasma/serum

Discussion of Structure

At the time we were confronted with this topic (1987) we had much experience with phytoactive substances, especially the whole area of Ginkgo Biloba. Regardless, the development of an acceptable sensitive method took us a period of more than two years (1987–1989). The LLOQ should be around 2–5 ng/ml plasma. There were 2 particularly problematic topics: Detection and sample preparation.

Concerning detection: UV at 285 nm was in our opinion not selective enough and just sensitive enough. Therefore, we tried to get a postcolumn derivatization for fluorescence detection. Precolumn was not possible because of 3 phenolic groups. All our experience with flavonols and flavonolglycosides or similar substances (e.g., polyphenols, catechins, procyanidines) was not helpful. We tried postcolumn derivatization with AlCl$_3$ or "Naturstoffreagens A", which under certain circumstances gave fluorescence derivates with specific polyphenols. Even using ECD for oxidation of the phenolic groups and different kinds of derivatization just behind the ECD did not produce success. ECD as a detector was not selective enough. So finally, only UV at 285 nm was an option. Therefore, a really good and selective sample preparation and enrichment was necessary. Finally, the extract of 100 μL plasma must be injected because of sensitivity reasons.

Sample Preparation

The solubility of Silibinin in water was poor and only good in acetone or different alcohols but not in more lipophilic solvents. Therefore, LLE was difficult to apply. Finally, ethyl acetate or diethyl ether was the only chance: in making 2 phases with water and being relatively hydrophilic; but when mixed with water or plasma an uptake of 1–3% water phase to ethyl acetate occurs. All more lipophilic solvents were not possible because of insufficient solubility for Silibinin. Only diethyl ether could also be a compromise. Also mixtures of different solvents were not suitable.

But by using ethyl acetate many endogenous substances occurred in the chromatogram disturbing the determination of Silibinin. What was the final solution: An extraction into diethyl ether at pH 8.5 (phenolic groups of Silibinin are not dissociated; all lipophilic acids stay in plasma because of pH [building salts]). This organic extract contains middle polar and nonpolar neutral substances, many middle polar and nonpolar bases and all middle polar phenols – and we were interested in all these.

After centrifugation and freezing out the plasma phase the organic phase was decanted and extracted again at pH 10.5–11 with a basic buffer. At this pH, all phenols go as phenolates into buffer. Also, all neutral substances and bases stay in the organic phase. This was the solution! The method and the results are published[37][38].

Useful tip 13: Enzymatic cleavage of glucuronides and sulfates

Numerous drugs – especially with phenolic and alcoholic functional groups – will be changed in the body to phase-2 metabolites-glucuronides and sulfates. Then they are eliminated easily through the kidney into the urine. Often, the orally administered drug is metabolized and only minor parts of this parent drug are eliminated through the urine. Therefore, for balancing – intake vs. elimination - especially phase-2 metabolites are important. As glucuronides or sulfates usually are not buyable, a good way is splitting off, e.g., the glucuronic acid and the determination of the parent drug. Two things are important from our point of view:

The unchanged parent drug might also be eliminated through the kidney. Therefore, determination of the parent drug before enzymatic cleavage is necessary. Unless you are only interested in a sum parameter. Some highly concentrated cheap mixtures of glucuronidase and sulfatase are on the market for cleavage. If you want to distinguish between glucuronides and sulfates – you can also buy the single enzyme.

As each of the enzymes (often also depending on the origin) has a pH optimum for cleavage please clear that up beforehand.

How much of the enzyme should I use? Please take only a small sample volume just enough for determination, so that you do not spoil the expensive enzymes. Please keep in mind that there are many glucuronides and sulfates in the urine – not only your drug. Therefore, use a great surplus of enzymes! For many of these enzymes body temperature (app. 37 °C) is the optimum temperature for cleavage. Do not waste time by using room temperature. Please consider that the chromatogram of a urine sample is quite different before and after cleavage – some peaks disappear and many peaks are new. Please check the selectivity of your method only with blank urine samples from different volunteers after cleavage. Sometimes, with cheap but highly active enzyme solutions you will get problems because of such a lot of peaks coming from the enzyme solution. Before fixing the cleavage conditions (consider temperature and especially time) it is necessary to take at least 2–3 real samples and split off for 1, 2, 4, 8 h to see the maximum cleavage yield.

We used many different enzymes with success (e.g., glucuronidase/sulfatase/glucosidase/lipase/protease).

For certain questions we even digested tissue samples completely with Subtilisin.

37 J. Liqu. Chromatogr. 17 (1994), 2777-2789 „Diastereomeric Separation of Free and Conjugated Silibinin in Plasma by Reversed Phase HPLC after Specific Extraction"

38 Int. J. Clin. Pharmacol. Ther. and Toxicol. 30 (1992), 134-138 „Study on dose-linearity of the pharmacokinetics of silibinin diastereomers using a new stereospecific assay"

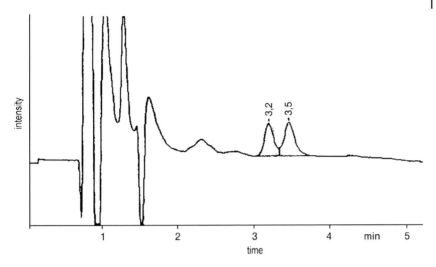

Chromatogram: 93 ng/ml as diastereomers (3.2' und 3.5') in plasma (calibration sample)

Useful tip 14: Special case: phenols and extraction (LLE)

We had to analyze during bioequivalence testing a few different studies for Silibinin over time. The validation and the first study worked brilliantly. For another study, we made the revalidation but we saw some outliers concerning calibration and QC samples – always too low. If you have a complicated method validated then you hand it over to the staff satisfied. You want to enjoy the fruits of this two-year development. And now such a mess! Now let us have a look at the staff to check if they have really worked precisely. Was the pH of the used buffers in the correct range (this is really essential for this method!) – but there was no deviation! Let us try it on my own with my whole experience! – and then again the same bad results. Obviously, this method does not work as properly as in former times. Who has to be blamed for this? In order not to stress you too long: A small detail was changed by the employee, without having been recognized as important. During the freezing process of plasma after extraction we took off the caps of the centrifuge tubes before freezing out (this is important for practical reasons).

As a difference to the starting conditions where everything worked well: There a small beaker (about 150 ml) was used for freezing out of about 6 samples together with acetone/dry ice. The new approach was a beaker of 250 ml for about 10–12 samples together. This beaker was higher in relation to the open sample tubes (about 80% higher). Which problem occurred? During the cooling process of the samples dry ice evaporated as CO_2 gas. This gas is relatively heavy and flows over the edge of the beaker (you can observe this). Sometimes, the CO_2 gas flew into the sample vials. By decanting the organic solvent this CO_2 gas changed the pH of the buffer from 10.5 to around 10 or even 9.5. Therefore, the re-extraction of Silibinin as phenolates did not work as well and was much less reproducible. As we had no internal standard (no similar natural substance can be used because of ubiquitous distribution in the nutrition) no compensation was possible when evaluating!

Rough Method Description of Our Determination (for the Free Part in Plasma)

To 1 ml plasma 0.2 ml 0.1 M Na_2HPO_4 solution was added (pH shifts to 8.5) and then extracted with 4 ml diethyl ether. After freezing and decanting the organic solvent the diethyl ether phase was re-extracted with 0.75 ml 0.1 M Na_2CO_3 solution. Part of the basic water phase was acidified with H_3PO_4 and than injected unto HPLC C18 column and chromatographed with UV detection at 285 nm.

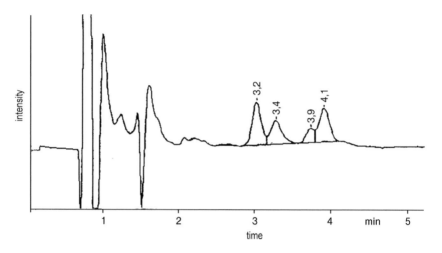

Chromatogram after oral application of a milk thistle extract: 103 ng/ml as diastereomers (3.2' und 3.4') in plasma (the following double peaks in the chromatogram are Isosilibinin-diastereomers (3.9' und 4.1'))

11.19
Valnemulin in Plasma, Different Tissue Types and in Animal Feed

Valnemulin is an antibiotic from the Novartis group used in animal health. It is a derivate of pleuromutilin.

MW: 564.8

C_{max}-value	protein binding (PB)	elimination half life $t_{1/2}$	pKa-value	Ionisation MS	Parent Ion
					Fragment Ion(s)
			4.3	pos	566

Matrix: Plasma/serum, different tissue, animal feed

Discussion of Structure

As you can see by observing the structure no chromophore can be found therefore UV and especially fluorescence detection is not possible with biological samples. Also, ECD is not useable. When we were confronted with this topic (1993) we saw a really good chance in derivatizing the primary amino group.

As we had derivatized many different substances with -NH2 using OPA, NDA or Fluram pre- and postcolumn in the past we focused on that. During tests we recognized that quantitative precolumn derivatization needs a different pH for that molecule comparing with, e.g., amino acids. Especially for derivatization in biological matrices the built fluorophore was not stable enough at pH around 9. Finally, we used precolumn NDA derivatization at pH 5 (instead of 9) and we could therefore avoid many problems[39][40].

39 Anal Chem 59 (1987) 1096-1101 "Naphthalene-2,3-dicarboxaldehyde/Cyanide Ion: A Rationally Designed Fluorogenic Reagent for Primary Amines"
40 LC-GC International 2 (1989) 20-27 "Naphthalenedialdehyde-cyanide: A versatile fluorogenic reagent for the LC analysis of peptides and other primary amines"

Rough Method Description of Our Determination

Derivatizing: After LLE from plasma and tissue samples and evaporation of the organic phase the residue was solved in 0.1 M KH_2PO_4 solution und KCN solution was added. After adding NDA reagent in acetonitrile the samples were kept for 15 min at 60 °C.

Plasma: To plasma samples Na_2CO_3 solution was added and then extracted with ethyl acetate.

The calibration range was 25–4812 ng/ml.

Tissue (muscle/kidney/liver/fat of pigs or cows): After tissue homogenization to a part of the samples there was given 5% Na_2CO_3 solution and detergents and extracted with ethyl acetate. The ethyl acetate phase was cleaned with 1 M KH_2PO_4 and then evaporated. The calibration range was 25–13 500 ng/g tissue.

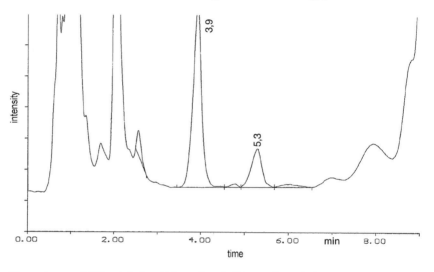

Chromatogram: 0.147 µg/g in liver, Valnemulin at 5.3', internal standard at 3.9'

Using animal feed, a mixture of acetonitrile and H_3PO_4 was used to get high recovery with an Ultraturrax. A small part was used after centrifugation by adding KH_2PO_4 solution to it and then derivatization.

After derivatization of samples of different origin, chromatography on C18 with 65% acetonitrile/35% 0.05 M methane sulfonic acid was done. Fluorescence detection with 420-nm excitation and 490-nm emission was used.

In the following decade when we also had to deal with metabolites or specific destruction products in animal feed we used MS and MS/MS also as detectors (we had MS from 1995 on and MS/MS from 1998 on).

One of the biggest challenges of this molecule for us was the recovery from different tissue types. Precolumn derivatization was checked within a short time and worked well for all samples. With LLE we had to find extraction conditions for each of 8 different tissue types. The conditions were partly dramatically different to get high recovery rates. As a summary we suppose different ionic binding to different tissue-specific molecules. One of the most general insights was that pH or specific acids and different detergents always play an important role.

Useful tip 15: Detergents for better recovery

By using LLE or even protein precipitation of distinct analytes from tissue – sometimes even from plasma or urine – the only way to obtain high recovery could be using detergents. By using nonionic detergents we often use Triton-X100 or different polyethylene glycols (which are complicated to eliminate).

Sometimes, amphoteric detergents like CHAPS were helpful. These types we like to use because they can be removed easily. By using detergents sometimes you hinder analytes to stick to different surfaces (or molecules, glass). Sometimes, you can make ion pairs (e.g., with dodecylsulfate) that become so lipophilic that you can extract them (with LLE or SPE).

Especially with tissue analytics in combination with analytes of complex structures with often basic groups detergents could be helpful, always higher recovery rates result. The concentration of detergents we use is in the millimolar range for the total sample solution. But as stressed before, try to get out the detergents before chromatography!

11.20
Vitamin B1 (Total Thiamine) in Plasma

Vitamin B1 is an endogenous substance as well as a pharmaceutical active substance. It is substituted for certain deficiencies.

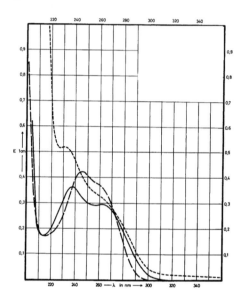

CI⁻

MW: 300.8

solvent symbol	methanol ———	water — · — · —	0.1M HCl — — — — —	0.1M NaOH · · · · · ·
absorption maximum	262 nm 238 nm		246 nm	232 nm
$E_{1cm}^{1\%}$	286 352		413	505
ε	9650 11 870		13 930	17 030

C$_{max}$-value	protein binding (PB)	elimination half life t$_{1/2}$	pKa-value	Ionisation MS	Parent Ion
					Fragment Ion(s)
200 mg oral → 30 ng/ml 2 – 13 ng/ml normal level			4.8	pos	265
					122/144/81

Matrix: Plasma/serum

Discussion of Structure

At the time we were confronted with this topic (1988) only some publications about endogenous plasma levels existed. But it was not clearly mentioned if they mean thiamine alone or together with the thiamine phosphates. Besides, there was no publication about plasma levels after oral application of vitamin B1. So, we had to take a basic decision, which analyte would fit to a bioequivalence study? Finally, after carefully checking all the literature we decided to determine thiamine and its monophosphate plus diphosphate as one parameter = thiamine. Therefore, quantitative cleavage of the two phosphates was necessary to determine one analyte = thiamine. This quantitative cleavage was possible after some testing with phosphatase.

The cited concentrations of endogenous levels after poor vitamin diet were rather low (around 5 ng/ml plasma). Therefore, UV detection was not possible. The volunteers should have a phase of vitamin-poor diet two days before oral application of a thiamine tablet. The day before application the same time points for blood drawing were used to see possible circadian effects.

From the literature we knew that thiamine could be oxidized to the fluorescent thiochrom. In the beginning we used coulometric ECD for oxidizing but finally we failed on the high NaOH concentration for thiochrom derivatization (pumped postcolumn). This NaOH blocked the cells of our ECD. So we used the postcolumn reaction with potassium-hexacyanoferrate und strong basic conditions to form thiochrom.

This reaction was extremely fast and reproducible. Within 3 min one chromatographic run including postcolumn derivatization was over. In the chromatogram we could also see the quality of phosphatase cleavage: Thiamine-mono- and -diphosphate gave a peak each in the chromatogram earlier than thiamine. This derivatization also works for the two phosphates.

Sample Preparation

This determination we published first in J. Pharm. Sci[41] and later we were asked to publish it also in Methods Enzymol[42].

The precision and reproducibility for the whole determination (including phosphatase cleavage) is excellent over the whole linearity range (<4% CV).

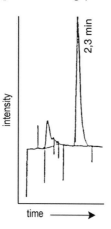

Chromatogram: 51 ng/ml in plasma (2.3')

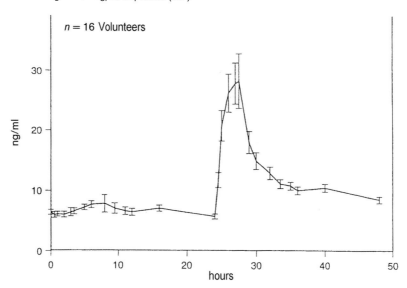

0–24 h: Native concentrations in plasma after usual nutrition (n = 16)
24–48 h: Plasma levels after oral vitamin application

41 J. Pharm. Sci. 82 (1993), 56-59 "Determination of Total Thiamine in Human Plasma for Oral Bioavailability Studies by HPLC with post column Derivatization"
42 Methods Enzymol. Vol. 280 (1997) "High-Performance Liquid Chromatography Determination of Total Thiamin in Human Plasma"

Table 11.1: Books, which I recommend for further information (the list also contains a few German books).

	Title	Publisher	Author	Important Topics
1.)	Hunnius Pharmazeutisches Wörterbuch	Walter de Gruyter 1993	Artur Burger, Helmut Wachter	Structures, chemical and physical constants of drugs and organic compounds
2.)	Plant Drug Analysis	Plant Drug Analysis 1984	H. Wagner, S. Bladt, E. M. Zgainski	Structures of exotic plant constituents
3.)	Handbook of Analytical Derivatization Reactions	Wiley-Interscience 1979	Daniel R. Knapp	Chemical derivatisation (but often only for GC)
4.) *	Detection-Oriented Derivatization Techniques in Liquid Chromatography	Marcel Dekker 1990	Henk Lingeman, Willy J. M. Underberg	Chemical derivatisation for HPLC
5.)	Chirality in Drug Research	Wiley-VCH 2006	Eric Francotte, Wolfgang Lindner	Chiral separation
6.)	Chromatographic Chiral Separations	Marcel Dekker 1988	Morris Zief, Laura J. Crane	Chiral separation
7.)	Elektrochemische Analytik	Springer-Verlag 1986	Günter Henze, Rolf Neeb	Electrochemical detection
8.) *	Biotransformation der Arzneimittel	Springer-Verlag 1990	Karl-Heinz Beyer	Metabolites and structures of these metabolites for a lot of drugs
9.) *	Biochemie für Mediziner und Naturwissenschaftler	Georg Thieme Verlag 1984	Peter Karlson	Endogenous substances

	Title	Publisher	Author	Important Topics
10.)	Handbook of Chemistry and Physics	CRC Press 1997	David R. Lide	Physical and chemical behaviour of solvents, melting and evaporation points, pKa-values, solubilities
11.)	The Pharmacological Basis of Therapeutics	McGraw-Hill 1995	Joel G. Hardman, Lee E. Limbird, Perry B. Molinoff, Raymond W. Ruddon, Alfred Goodman Gilman	Structures, pharmacokinetic, metabolism
12.)	Bioanalytik	Spektrum Akademischer Verlag 1998	F. Lottspeich, H. Zorbas	Lipids, peptides, proteins
13.) *	Clarke's Isolation and Identification of Drugs	The Pharmaceutical Press 1986	A.C. Moffat, J.V. Jackson, M.S. Moss, B. Widdop	Lot of information for drugs: Pharmacokinetic, methods for determination, protein binding, therapeutic concentrations
14.) *	Martindale – The Complete Drug Reference	Pharmaceutical Press 2005	Sean C. Sweetman	Lot of information for drugs: Pharmacokinetic, dose, metabolites, protein binding (very important)
15.) **	The Merck Index	Merck & Co 2006	Maryadele J. O'Neil	Structure for almost all drugs, solubility, molecular weight, pKa-values, formula index (very important)
16.) *	Römpp Chemielexikon	Georg Thieme Verlag 1995	Jürgen Falbe, Manfred Regitz	Helpful for overview of substance classes

	Title	Publisher	Author	Important Topics
17.)	Diagnostic Samples: From the Patient to the Laboratory	Wiley-Blackwell 2009	Walter G. Guder, Sheshadri Narayanan, Hermann Wisser, Bernd Zawta	Gaining blood samples, storage, therapeutic drug monitoring
18.)*	Wissenschaftlich Tabellen Geigy 1–6	Ciba-Geigy 1977		Concentration of a lot of endogenous substances (plasma, urine, saliva ...)

* very important to me

Table 11.2: Properties of substances

name	C_{max}-value	protein binding (PB)	elimination half life $t_{1/2}$	pKa-value	Ionisation MS	Parent Ion / Fragment Ion(s)
Acetylcarnitine	native value ca. 1000 ng/ml			-1.8	pos	204 / 85
Acetylcysteine	200 mg oral → 1.5 µg/ml	78 %	6 h	9.5	neg	162 / 84/33
Acyclovir	200 mg oral → 400 ng/ml (usual 0.5 – 15 µg/ml)	9 – 33 %	1.5 – 6 h	-1.7	pos	226 / 152/135/110
Allopurinol	100 mg oral → 0.5 µg/ml (usual 1 – 5 µg/ml)	<5 %	0.5 – 2 h Oxipurnol 18 – 43 h	9.4	neg	135 / 64/92
Ambroxol	30 mg oral → 80 ng/ml	80 – 90 %	7 – 12 h	3.1	pos	379 / 264/104/116
Amiloride	5 mg oral → 15 ng/ml	40 %	10 – 20 h	8.7	pos	230 / 171/116/143
Amoxicillin	500 mg oral → 8 µg/ml	20 %	1 h	2.4/7.4/9.6	pos	366 / 349/208/134/114
Amphetamine	10 mg oral → 20 ng/ml after 5 mg Selegelin oral → 1.5 ng/ml	15 – 40 %	4 – 8 h	9.9	pos	136 / 91/65/119
Ampicillin	500 mg oral → 3 µg/ml	20 %	1 – 2 h	2.5/7.3	pos	350 / 106/192
Bacitracin		40 – 90 %			pos (2++)	712 / 119

name	C_{max}-value	protein binding (PB)	elimination half life t_{1/2}	pKa-value	Ionisation MS	Parent Ion / Fragment Ion(s)
Bendroflumethiazide	2.5 mg oral	high	3 – 4 h	1.8	pos	422 / 122/405
					neg	420 / 328/289/224
Benzbromarone	100 mg oral → 3 µg/ml	high	3 h	6.0	pos	423/425/427
Benzoic Acid				4.2	pos	123 / 61/81/40
Bezafibrate	200 mg oral → 7 µg/ml	95 %	2 h	4.3	pos	362 / 316/121/139
					neg	360 / 274/154
Budesonide	250 µg spray → 280 pg/ml	88 %	2.8 h		pos	431 / 413/147/173
					neg	489 / 357
Butorphanol	15 mg oral → 2 ng/ml	80 %	2 – 4 h	3.7	pos	328 / 310
Caffeine	100 mg oral → 3 µg/ml (usual 2 – 10 µg/ml)	35 %	2 – 10 h	14.0	pos	195 / 138/110/83
Canrenone	after 100 mg oral Spironolacton → 350 ng/ml Canrenone (usual 50 – 250 ng/ml)	>90 %	15 – 21 h		pos	341 / 205/323
					neg	357 / 313/339/145
Captopril	100 mg oral → 0.8 µg/ml	30 %	1 – 2 h	3.7/9.8	pos	218 / 116/70/75
					neg	216 / 182/114

name	C_{max}-value	protein binding (PB)	elimination half life $t_{1/2}$	pKa-value	Ionisation MS	Parent Ion / Fragment Ion(s)
Carbamazepine	200 mg oral → 4 µg/ml (usual 3 – 12 µg/ml)	75 %	10 – 30 h		pos	237 / 194/192/179
Carvedilol	20 mg oral → 50 ng/ml	98 %	4 – 8 h	3.0	pos	407 / 222/100/194
Chlorthalidone (Blood)	25 mg oral → 1.5 µg/ml whole blood (ca. 98 % erythrocytes)	75 % plasma	35 – 70 h	9.4	pos	339 / 322/243
Ciclesonide	after inhalation >100 pg/ml	>99 %	0.7 h (3.5 h Des-CIC)		neg	599 (acetic adduct) / 339
Cimetidine	400 mg oral → 2 µg/ml (usual 0.5 – 1 µg/ml)	13 – 26 %	1 – 3 h	6.8	pos / neg	253 / 159/95/117 ; 251 / 97/123
Cinnarizine	25 mg oral → 100 ng/ml		5 h	5.4	pos	369 / 167/152
Clarithromycin	250 mg oral → 900 ng/ml	42 – 50 %	3 – 4 h	3.2	pos	748 / 590/158
Codeine	50 mg oral → 0.2 µg/ml (usual 0.025 – 0.15 µg/ml)	7 – 25 %	2 – 4 h	8.2	pos	300 / 165/153
Dextrorphan	after 60 mg Dextrometorphan → 700 ng/ml		3 h	3.4	pos	258 / 199
Diazepam	5 mg oral → 100 ng/ml (usual 200 – 500 ng/ml)	98 – 99 %	20 – 100 h	3.3	pos	285 / 193/154/222

name	C_{max}-value	protein binding (PB)	elimination half life $t_{1/2}$	pKa-value	Ionisation MS	Parent Ion / Fragment Ion(s)
Diclofenac	50 mg oral → 800 ng/ml (usual 50 – 2500 ng/ml)	>99 %	1 – 2 h	3.1	pos	296 / 214/250/178
Dihydralazine	25 mg oral → 50 ng/ml		1 – 2 h	0.2	pos	191 / 129/102/174
Diphenhydramine	50 mg oral → 50 ng/ml (usual 80 – 400 ng/ml)	78 %	2 – 9 h	3.1	pos	256 / 165/152/167
Dipyridamole	100 mg oral → 2 µg/ml	>90 %	12 h	6.4	pos	505 / 429/385
Doxorubicin	1.5 mg/kg iv		3 and 40 h		pos	544 / 361/321/397
					neg	542 / 395
Doxycycline	100 mg oral → 3 µg/ml (usual 3 – 5.3 µg/ml)	82 – 90 %	22 h	3.5/7.7/ 9.5	pos	446 (decomp.) / 156/428
Duramycin	>1 – 10 ng/ml				pos (2^{++})	1007 / 1007
Erythromycin	500 mg oral → 2 µg/ml	70 – 80 %	1 – 3 h	8.9	pos	734 / 576/158/83
Fluticasone Propionate	>50 – 100 pg/ml after inhalation	81 – 95 %	3.5 h		pos	501 / 313/293/205
Formoterol	after inhalation of 36 µg → 30 pg/ml		8 h	1.4	pos	345 / 149

name	C_{max}-value	protein binding (PB)	elimination half life $t_{1/2}$	pKa-value	Ionisation MS	Parent Ion / Fragment Ion(s)
Furosemide	80 mg oral → 2.3 µg/ml (usual 2 – 5 µg/ml)	97 %	1 – 3 h	3.9	pos	331 / 119/303/299
					neg	329 / 285/205/78
Glibenclamide	3.5 mg oral → 150 ng/ml (usual maximum 100 – 300 ng/ml)	99 %	5 – 16 h	5.3	pos	494 / 369/169/304
Hydrochlorthiazide	50 mg oral → 300 ng/ml (usual 50 – 160 ng/ml)	60 %	2.5 – 10 h	7.0/9.2	neg	296 / 205/269/78
Ibuprofen	200 mg oral → 20 µg/ml (usual 15 – 30 µg/ml)	99 %	2 h	3.8	neg	206 / 161/91/119
Indomethazin	50 mg oral → 1.5 µg/ml (usual 0.8 – 2.5 µg/ml)	90 – 99 %	3 – 15 h	4.5	pos	358 / 139
Isosorbid-5-mononitrate	20 mg oral → 600 ng/ml	>5 %	3 – 7 h		pos	192 /
Itraconazol	200 mg daily → 2 µg/ml (usual 0.2 – 2 µg/ml)	99.8 %	20 h	6.2	pos	705 / 392/432/450
Ketoconazole	400 mg oral → 6.5 µg/ml	99 %	6 – 10 h	4.5	pos	531 / 489/148/219
Lansoprazole	30 mg oral → 900 µg/ml	97 %	1 – 2 h	2.8	neg	368 / 164

name	C_{max}-value	protein binding (PB)	elimination half life $t_{1/2}$	pKa-value	Ionisation MS	Parent Ion / Fragment Ion(s)
Levodopa	250 mg oral → 1400 ng/ml (usual 200 – 2500 ng/ml)	<5 %	1 – 3 h	2.3/8.7/ 9.7/13.4	pos / neg	198 / 152/139/181 ; 196 / 135/109
Lornoxicam	4 mg oral → 2 µg/ml	99 %	3 – 4 h	3.0	pos / neg	372 / 121 ; 370 / 306/186/171/150
Mefenamin Acid	1000 mg → 10 µg/ml	99 %	3 – 4 h	4.2	pos / neg	242 / 224/209/180 ; 240 / 196/180
Metformin	500 mg oral → 1.6 µg/ml (usual 0.1 – 1.3 µg/ml)	<5 %	3 – 6 h	2.8/11.5	pos	130 / 60/71/85
5-Methoxypsoralen + 8-Methoxypsoralen	40 mg oral → 50 ng/ml ; 30 mg oral → 300 ng/ml	>90 %	2.3 h ; 3 h		pos	217 / 174
Metoprolol	100 mg oral → 150 ng/ml (usual 100 – 600 ng/ml)	12 %	2 – 7 h	9.7	pos	268 / 77/116/103
Metronidazole	250 mg oral → 9 µg/ml (usual 10 – 30 µg/ml)	11 %	8 h	2.5	pos	172 / 128/82
Midazolam	20 mg oral → 260 ng/ml (usual 80 – 250 ng/ml)	95 %	2 h	4.3	pos	326 / 291/244/249

name	C_{max}-value	protein binding (PB)	elimination half life $t_{1/2}$	pKa-value	Ionisation MS	Parent Ion Fragment Ion(s)
Minocycline	50 mg oral → 1.5 µg/ml	75 %	11 – 26 h	2.8/5.0/ 7.8/9.5	pos	458 441/352/283
Montelukast	10 mg oral → 400 ng/ml	>99 %	2.7 – 5.5 h	9.5	pos	586 568
Nabilone	2 mg oral → 2 ng/ml		2 h	7.1	pos	373 247
Neomycin	2 mg oral → 2 ng/ml		4 h		pos (2++)	308 454
Nicorandil	20 mg oral → 250 ng/ml	few	1 h	0.4	pos	212 136
Nifedipine	10 mg oral → 80 ng/ml (usual 10 – 200 ng/ml)	96 %	2 – 6 h	2.5	pos	347 254/195/167
Nitrofurantoin	100 mg oral → 1 µg/ml	60 %	0.5 – 1 h	7.2	neg	237
Norfloxacin	400 mg oral → 1.3 µg/ml	14 %	3 – 4 h		pos	320 302/233/276
Omeprazole	20 mg oral → 480 ng/ml (usual maximum 800 – 4400 ng/ml)	95 %	0.5 – 3 h	2.2	pos	345 297/149/282
Oxazepam	40 mg oral → 600 ng/ml (usual 0.5 – 2.0 µg/ml)	95 %	4 – 25 h	1.7/11.6	pos	287 241/269/104
Paclitaxel	50 – 500 ng/ml	95 – 98 %	3 – 50 h	3.3	pos	854 286

name	C$_{max}$-value	protein binding (PB)	elimination half life t$_{1/2}$	pKa-value	Ionisation MS	Parent Ion / Fragment Ion(s)
Pantoprazole	40 mg oral → 1.2 μg/ml (usual maximum 1.1 – 3.1 μg/ml)	98 %	1 h	1.7	pos	384 / 200
Paracetamol	10 – 20 μg/ml (usual therapeutic range toxic/poisonous >300 μg/ml) 500 mg oral → 5 μg/ml	<5 %	1.5 – 3 h	9.5	pos	152 / 110/65/93
Paroxetine	25 mg oral → 15 ng/ml (usual 10 – 100 ng/ml)	95 %	21 h	4.7	pos	330 / 192/70/135
Penicillin V	700 mg oral → 7 μg/ml	80 %	0.5 h	2.7	pos / neg	351 / 160/114 / 349 / 93/114
Pentoxifylline	600 mg oral retard → 100 ng/ml (usual 20 – 200 ng/ml)	<5 %	0.4 – 0.8 h	0.6	pos	279 / 181/99/138
Phenobarbital	150 mg oral → 10 – 30 μg/ml (usual 10 – 40 μg/ml)	50 %	50 – 150 h	7.4	neg	231 / 144/188
Phenylbutazone	300 mg oral → 38 μg/ml (usual 50 – 150 μg/ml)	99 %	28 – 120 h	4.4	pos	309 / 160/77/92

name	C_{max}-value	protein binding (PB)	elimination half life $t_{1/2}$	pKa-value	Ionisation MS	Parent Ion	Fragment Ion(s)
Phenytoin	100 mg oral → 2.2 µg/ml → (usual 10 – 20 µg/ml)	90 %	7 – 60 h	8.3	neg neg	253 251	182/104 102/208
Pimelic Acid	usual 2 – 10 ng/ml				neg	159	97
Piracetam	800 mg oral → 20 µg/ml	15 %	5.2 h		pos	143	126/98/70
Pirenzepine	50 mg oral → 40 ng/ml	10 %	11 h	2.1/8.1	pos	352	113/70/252
Piroxicam	20 mg oral → 2 µg/ml (usual 2 – 20 µg/ml)	99 %	30 – 60 h	2.6	pos neg	332 330	95/164/121 146/131/119
8-Prenylnaringenin					pos	341	165/285
Propafenone	(usual 250 – 1000 ng/ml)	85 – 95 %	5 bzw. 20 h	3.4	pos	342	116/72/324
Propranolol	40 mg oral → 30 ng/ml >20 ng/ml active (usual 50 – 300 ng/ml)	90 %	2 – 6 h	9.5	pos	260	116/183/155
Propyphenazone	300 – 1500 mg daily (usual 1.5 – 3.5 µg/ml)	10 %	1 – 1.5 h	2.0	pos	231	189/56/201/77

name	C$_{max}$-value	protein binding (PB)	elimination half life t$_{1/2}$	pKa-value	Ionisation MS	Parent Ion / Fragment Ion(s)
Roflumilast	0.5 mg oral → 7 ng/ml	99 %	10 h	3.5	pos	403 / 187
Salbutamol	200 µg inhal. → 500 pg/ml (usual 1 – 20 ng/ml)	low	2 – 7 h	9.3/10.3	pos	240 / 148/166/222
Salicylic Acid	therapeutic 50 – 300 µg/ml	50 – 90 %	2 – 4 h	3.0/13.4	neg	137 / 93/65/75
Salmeterol	50 – 200 pg/ml	93 %		4.2	pos	416 / 398/232/91
Silibinin = Silymarin	100 mg oral → 110 ng/ml free / total 2500 ng/ml (Glucuronid)		2.5 h	1.9	neg	481 / 301/125
Temazepam	20 mg oral → 700 ng/ml (usual 200 – 800 ng/ml)	97 %	3 – 38 h	1.6	pos	301 / 225/177/193
Theophylline	(usual 8 – 20 µg/ml)	56 %	3 – 13 h	<1/8.6	pos	181 / 124/81/69
Tolperisone (+Enantiomers)	300 mg oral → 300 ng/ml	90 %	2.5 h	3.7	pos	246 / 55
Tramadol	50 mg retard → 70 ng/ml (usual 300 – 900 ng/ml)	<5 %	6 h	8.3	pos	264 / 58/42
Triamterene	50 mg oral → 60 ng/ml	45 – 70 %	2 – 4 h	6.2	pos	254 / 237/104/195
Tryptophan	usual up to 16 µg/ml			2.4/9.4	pos	205 / 188/146/118

name	C$_{max}$-value	protein binding (PB)	elimination half life t$_{1/2}$	pKa-value	Ionisation MS	Parent Ion / Fragment Ion(s)
Valnemulin				4.3	pos	566 / —
Verapamil	40 mg oral → 60 ng/ml >100 ng/ml active (usual 50 – 750 ng/ml)	90 %	2 – 7 h	4.8	pos	455 / 165/150/303
Vitamin B 1 (Thiamine)	200 mg oral → 30 ng/ml 2 – 13 ng/ml normal level			4.8	pos	265 / 122/144/81
Vitamin B 2 (Riboflavin)	normal level 30 – 150 ng/ml				pos	377 / 228/198/170
Vitamin B 6 (Pyridoxal)	40 mg oral → 200 ng/ml 5 – 24 ng/ml normal level					170 / 134/152/77
Vitamin E (Tocopherol)	5 – 20 µg/ml			12.2	pos	431 / 165

Appendix

Short Description of Determination for about 100 Substances

Here you can find the most important information about 100 substances often used in clinical analysis. Most of these determinations are self-approved.

You can find in clear form not only molecule structures but also UV spectra, rough MS data and published bioanalytical methods (our own and published) with determination limits. Furthermore, remarks on oral dosages and resulting C_{max} values, pKa values, remarks on protein binding (PB) and elimination half-life ($t_{1/2}$) are listed. But care should be taken, all remarks are rough estimations (sometimes a mixture of different citations) and not suitable for citation! These data are benchmarks to get a feeling for that substance, in particular to know the concentration range your method should reach. The oral dosages and the resulting C_{max} values are rough estimations of cited mean concentrations from different bioavailability studies, cited literature or our own results from studies. Sometimes dosages are low, sometimes high, some tablets for instance should be applied four times a day, some only once a day (partly no matter concerning half-life of the active ingredient). Also, the elimination half-life is a mixture of cited results – sometimes the half-life of distribution phase, sometimes the terminal phase. With some applied methods it is not possible to measure the rather low concentrations of specific substances in the terminal phase. The data differ partly considerably: In one case we found in the literature from different sources half-lives of 2–30 h. Sometimes these are results from volunteers and sometimes from sick people. Even pKa values sometimes were variable in literature depending on the source. Exemplarily, we want to show you how helpful those data can be:

Example: Amoxicillin

C_{max}-value	protein binding (PB)	elimination half life $t_{1/2}$	pKa-value	Ionisation MS	Parent Ion
					Fragment Ion(s)
500 mg oral → 8 µg/ml	20 %	1 h	2.4/7.4/ 9.6	pos	366
					349/208/134/ 114

a.) Information: 500 mg oral results in a C_{max} of 8 µg/ml plasma.

Insight: A method for checking compliance (if necessary for such a drug) should be able to measure 1/10 of C_{max} – this means 0.8 µg/ml. As a consequence you have to inject by using UV detection 1 µL plasma (or the extract of 1 µL) onto the column. This should not be a problem!

b.) Information: The elimination half-life is around 1 h.

Insight: A lot of drugs have the C_{max} value after 1–3 h after oral ingestion. We didn't want to make more precise remarks because of the huge variability: depending on the galenic formulation, depending on the patient, depending on being sober or after a meal (also depending on high or low fat content of the food).

In this case we know that 3–5 h after (scheduled!) ingestion you should find the C_{max} value or for instance a value of four times lower the half-life: for example C_{max} at 1 h blood drawing after 5 h = 4 x $t_{1/2}$ (= 1 h for amoxicillin). This means at 8 µg/ml at last (4/2/1/0.5) 0.5 µg/ml. If we compare this value with the scheduled determination limit of 0.8 µg/ml, then you really could be stressed! Either you look more closely at the pharmacokinetic profile of the used galenic formulation (you should get this from the distributor) or you ask the medical doctor to draw blood after about 2 and 4 h after ingestion. The other chance is to develop a more sensitive method down to 0.1 µg/ml. By having a half-life of 20 h for a given substance it would be enough to await the absorption time and then do blood drawing within 10–20 h after ingestion.

c.) Information: Protein binding (PB) is 20%.

Insight: The cited levels concerning protein binding can help you to avoid some mistakes during sample preparation. For amoxicillin (PB 20%) you can't use protein precipitation with acids (e.g. TCA or $HClO_4$). In this case you would lose these 20%. If the protein binding is very similar from patient to patient you could even accept the loss of 20%. But there are a lot of drugs with a high variability in protein binding from patient to patient. If a molecule has a protein binding of 80–100% it is senseless to use protein precipitation with acids. For highly protein bound molecules often a protein precipitation with acetonitrile, methanol, ethanol or acetone is fitting (getting high recovery).

d.) Information: pKa value 2.4/7.4/9.6

Insight: What could be the message for an analyst? You can see that you have acidic or basic groups in the molecule (please check carefully the molecular structure!). This knowledge should influence your thoughts concerning sample preparation and HPLC analysis. From the pKa values you know when a substance is undissociated or dissociated. This influences the solubility in different solvents dramatically and also the chromatographic behavior under different pH conditions. In addition, it could be helpful through pH change to influence the protein binding and also the recovery in relation to plasma or tissue proteins. Substances with two or more pKa values are not seldom amphoteric (please look at the molecular structure!). Concerning substances with four pKa values like Tetracycline you should see a lot of problems coming!

Amoxicillin has three pKa values: 2.4 (-COOH)/7.4 (-NH$_2$)/9.6 (-phenolic group). The stability of such a molecule often depends on the pH. Chromatography on RP needs undissociated molecules or ion pairs. In this specific case you could stay below pH 9 (because of the phenolic group), between pH 8–9 because of the amino group or as an ion pair with –COOH higher pH 3–4. The reality often looks a little different, but as a rough rule the pKa values are helpful for chromatography. Also for LLE the pKa values are important to determine: You can only extract (enough lipophilic) molecules undissociated into organic (hydrophobic) solvents.

Dissociated molecules even they are lipophilic solve in the water phase or buffer.

It is a pity but we could not find pKa values for all substances in the literature. No pKa value cited does not necessary mean that the substance has no functional group in this direction. Prof. Ulrich Jordis (TU-Vienna, Institute of Applied Synthesis) supported me with a lot of missing pKa values from the literature.

MS-Information

e.) Ionisation MS: either pos = positiv or neg = negativ.

The ionisation mode used for these citations (often found in literature even own experience) was gained by ESI or APCI. For ionic substances often ESI was applied, for lipophilic, neutral substances often APCI was applied.

f.) Parent Ion and Fragment Ion(s): The parent ion is usually the quasi molecular ion (H$^+$ or H$^-$), sometimes an adduct ion or an ion after loss of water or ammonia.

The fragment ions usually were gained by MRM in tandem mass spectrometers, seldom by ion trap systems. The fragment ions are ordered by signal intensity. They can also occur in single quad systems using collision induced dissociation in the ion source (use of higher voltage).

The determination of substances listed in the appendix shows our solutions during the last 30 years, as well as the last level in analytics cited in Analytical Abstracts or the original literature.

During a couple of years we especially used systems from AB Sciex (API 3000/API 4000/API 5000) for quantitation and from Thermo Scientific (LCQ Deca and LTQ-Orbitrap XL) for structural elucidation. These systems are not the ones a lot of analysts use and therefore they can not solve their analytical questions in this way. Therefore HPLC-UV, HPLC-ECD, HPLC fluorescence or HPLC fluorescence after derivatization could be helpful for your routine work.

The whole range of antiasthmatic drugs, however, forms an exception. Applying a maximum of 500 μg of substance in an inhalative application leads to such low plasma levels that you need the most sensitive MS/MS systems that are nowadays on the market. Furthermore, substances with no chromophore and functional group for derivatization are also an exception if they are applied lower than 10–50 mg per person orally. Also, substances with a very high first pass can lead to such low levels of parent drug in plasma that you don't have any chance without MS/MS.

But one point should still be kept in mind: There are some volatile substance classes where GC with FID, MS or MS/MS is the method of choice. TLC is a good method but for quantitation or real sensitive determination HPLC is much better. Concerning side products or even sometimes with metabolites TLC could be a good option. After separation (all substances between Rf O-1) and drying the plate you can choose a lot of derivatization agents for specific functional groups. Sometimes, TLC can also be a complementary method.

Substances listed in the Appendix

Only determination in plasma/serum is cited. For protein precipitation usually acetonitrile was used but sometimes also methanol, ethanol or a strong acid (e.g. TCA, $HClO_4$). If there are rough estimations-increments of 10. The sign "<" means bad or very weak.

A	Acetylcarnitine		Cinnarizine
	Acetylcysteine		Clarithromycin
	Acyclovir		Codeine
	Allopurinol		
	Ambroxol	D	Dextrorphan
	Amiloride		Diazepam
	Amoxicillin		Diclofenac
	Amphetamine		Dihydralazine
	Ampicillin		Diphenhydramine
			Dipyridamole
B	Bacitracin		Doxorubicin
	Bendroflumethiazide		Doxycycline
	Benzbromarone		Duramycin
	Benzoic Acid		
	Bezafibrate	E	Erythromycin
	Budesonide		
	Butorphanol	F	Fluticasone Propionate
			Formoterol
C	Caffeine		Furosemide
	Canrenone		
	Captopril	G	Glibenclamide
	Carbamazepine		
	Carvedilol	H	Hydrochlorthiazide
	Chlorthalidone (Blood)		
	Ciclesonide	I	Ibuprofen
	Cimetidine		Indomethacin

	Isosorbide-5-mononitrate		Pentoxyphylline
	Itraconazole		Phenobarbital
			Phenylbutazone
K	Ketoconazole		Phenytoin
			Pimelic Acid
L	Lansoprazole		Piracetam
	Levodopa		Pirenzepine
	Lornoxicam		Piroxicam
			8-Prenylnaringenin
M	Mefenamic Acid		Propafenone
	Metformin		Propranolol
	Metoprolol		Propyphenazone
	5-Methoxypsoralen		
	8-Methoxypsoralen	**R**	Roflumilast
	Metronidazole		
	Midazolam	**S**	Salbutamol
	Minocycline		Salicylic Acid
	Montelukast		Salmeterol
			Silibinin
N	Nabilone		
	Neomycin	**T**	Temazepam
	Nicorandil		Theophylline
	Nifedipine		Tolperisone (+ Enantiomers)
	Nitrofurantoin		Tramadol
	Norfloxacin		Triamterene
			Tryptophan
O	Omeprazole		
	Oxazepam	**V**	Valnemulin
			Verapamil
P	Paclitaxel		Vitamin B1 (Thiamine)
	Pantoprazole		Vitamin B2 (Riboflavin)
	Paracetamol		Vitamin B6 (Pyridoxal)
	Paroxetine		Vitamin E (Tocopherol)
	Penicillin V		

Short Explanation of Tables Presented in the Appendix

1. Every estimation of amount or concentration needs to be done in steps of 10.
2. < means very bad
3. Precipitation: mostly acetonitrile, sometimes methanol, in few cases $HClO_4$ or TCA
4. Sources for the MS-data:
   a) Own spectra
   b) T+K (2008) 75, 149 Guessregen et al.
   c) Anal. Bioanal. Chem (2009) 395, 2521 Dresen et al.
   d) Some specific literature
Indications always refer to plasma or serum.

Acetylcarnitine

MW: 203.2

no UV
no E1/1

C$_{max}$-value	protein binding (PB)	elimination half life t$_{1/2}$	pKa-value	Ionisation MS	Parent Ion
					Fragment Ion(s)
native value ca. 1000 ng/ml			-1.8	pos	204
					85

detection limits	LOD	precipitation	LLE	SPE
UV	<			44 ng/ml [2]
FL	<			
ECD	<			
MS/MS	10 pg	10 ng/ml [1]		

[1] Chemical Monthly 136 (2005) 1425-1442; precipitation HPLC-MS/MS 10 ng/ml
[2] J Liqu Chromatogr Relat Technol 24 (2001), 555-563; SPE HPLC-UV (260 nm) after derivatization 44 ng/ml

Acetylcysteine

MW: 163.2

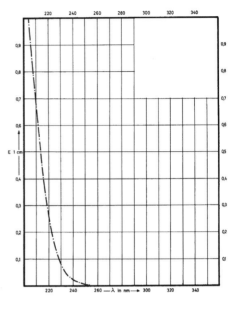

solvent symbol	methanol ———	water —·—·—	0.1M HCl —————	0.1M NaOH ······
absorption maximum				
$E_{1cm}^{1\%}$				
ε				

C_{max}-value	protein binding (PB)	elimination half life $t_{1/2}$	pKa-value	Ionisation MS	Parent Ion ——— Fragment Ion(s)
200 mg oral → 1.5 µg/ml	78 %	6 h	9.5	neg	162 ——— 84/33

detection limits	LOD	precipitation	LLE	SPE
UV	10 ng (220 nm)			
FL	<	1 ng/ml [1, 2]		
ECD	<			
MS/MS	10 pg			

(1) Biopharm Drug Disposit 12 (1991) 343-353; HPLC-Fluor (325/475 nm) after HClO$_4$-precipitation derivatization 1 ng/ml
(2) Biomed Chromatogr 20 (2006) 415-422; HPLC-Fluor (330/376 nm) after derivatization 4 ng/ml

Acyclovir

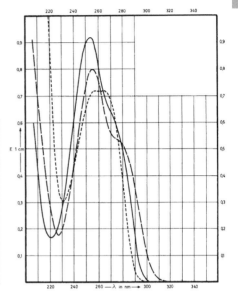

MW: 225.2

solvent symbol	methanol ———	water —·—·—	0.1M HCl —————	0.1M NaOH ······
absorption maximum	254 nm		255 nm	257 nm 264 nm
$E_{1cm}^{1\%}$	611		530	475 475
ε	13800		11900	10700 10700

C_{max}-value	protein binding (PB)	elimination half life $t_{1/2}$	pKa-value	Ionisation MS	Parent Ion ——— Fragment Ion(s)
200 mg oral → 400 ng/ml (usual 0.5 – 15 µg/ml)	9 – 33 %	1.5 – 6 h	-1.7	pos	226 ——— 152/135/110

detection limits	LOD	precipitation	LLE	SPE
UV	1 ng (254 nm)	63 ng/ml [4]		
FL	10 pg [1]	10 ng/ml [1]	10 ng/ml [3]	
ECD	<			
MS/MS		2 ng/ml [2]		

(1) J Chromatogr 583 (1992) 122-127; HClO4-precipitation 10 ng/ml HPLC-Fluor (260/375 nm)
(2) Biomed Chromatogr 23 (2009) 132-140; precipitation; HPLC-MS/MS 2 ng/ml
(3) J Chromatogr B 816 (2005) 327-331, extraction; HPLC 10 ng/ml
(4) Biomed Chromatogr 17 (2003) 500-503, precipitation; HPLC-UV (254 nm) 63 ng/ml

Allopurinol

MW: 136.1

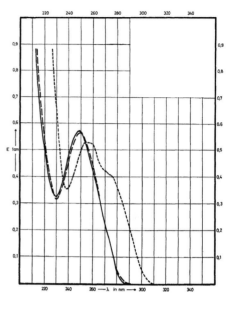

solvent symbol	methanol ———	water —·—·—	0.1M HCl ————	0.1M NaOH ······
absorption maximum	249 nm		250 nm	255 nm
$E_{1cm}^{1\%}$	566		555	521
ε	7700		7550	7900

C_{max}-value	protein binding (PB)	elimination half life $t_{1/2}$	pKa-value	Ionisation MS	Parent Ion ——— Fragment Ion(s)
100 mg oral → 0.5 µg/ml (usual 1 – 5 µg/ml	<5 %	0.5 – 2 h Oxipurnol 18 – 43 h	9.4	neg	135 ——— 64/92

detection limits	LOD	precipitation	LLE	SPE
UV	1 ng (249 nm)	50 ng/ml [1]		
FL				
ECD				6 ng/ml [2]
MS/MS		10 ng/ml [3]		

(1) Drug Res 30 (1980) 1855-1857; TCA-precipitation HPLC-UV (250 nm) 50 ng/ml
(2) Pharm Res 8 (1991) 653-655; SPE, HPLC-ECD 6 ng/ml
(3) HPLC-MS/MS

Ambroxol

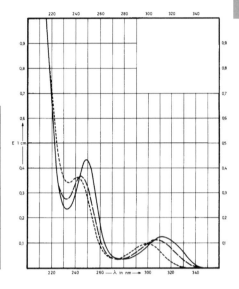

MW: 378.1

solvent symbol	methanol ———	water —·—·—	0.1M HCl —————	0.1M NaOH ······
absorption maximum	313 nm 248 nm		307 nm 244 nm	299 nm 241 nm
$E_{1cm}^{1\%}$	79 282		72 241	64 240
ε	3300 11700		3000 10000	2700 9900

C_{max}-value	protein binding (PB)	elimination half life $t_{1/2}$	pKa-value	Ionisation MS	Parent Ion ——— Fragment Ion(s)
30 mg oral → 80 ng/ml	80 – 90 %	7 – 12 h	3.1	pos	379 ——— 264/104/116

detection limits	LOD	precipitation	LLE	SPE
UV	1 ng (248 nm)		5 ng/ml [2]	
FL				
ECD	0.1 ng		1 ng/ml [3]	
MS/MS		1 ng/ml [1]		

[1] HPLC-MS/MS 1 ng/ml ACN-precipitation
[2] J Chromatogr 421 (1987) 211-215; extraction HPLC-UV (242 nm) 5 ng/ml
[3] J Chromatogr 490 (1989) 464-469; extraction HPLC-ECD 1 ng/ml

Amiloride

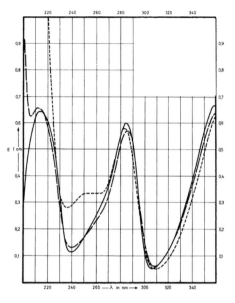

MW: 229.6

solvent symbol	methanol ———	water —·—·—	0.1M HCl —————	0.1M NaOH ······
absorption maximum	361 nm 286 nm		361 nm 286 nm	366 nm 284 nm
$E_{1cm}^{1\%}$	738 660		705 630	725 635
ε	19640 17560		18760 16760	19290 16900

* Referred to dried

C_{max}-value	protein binding (PB)	elimination half life $t_{1/2}$	pKa-value	Ionisation MS	Parent Ion / Fragment Ion(s)
5 mg oral → 15 ng/ml	40 %	10 – 20 h	8.7	pos	230 / 171/116/143

detection limits	LOD	precipitation	LLE	SPE
UV	0.1 ng (361 nm)		0.7 ng/ml [1] / 0.2 ng/ml [2]	
FL	10 pg	0.5 ng/ml [3]	0.5 ng/ml [4]	
ECD				
MS/MS				

[1] Therapy week 36 (1985) 56-60; extraction HPLC-UV (315 nm) 0.7 ng/ml
[2] J Liqu Chromatogr Relat Technol 31 (2008) 2455-2466, extraction; HPLC-UV (360 nm) 0.2 ng/ml
[3] J Chromatogr 377 (1986) 399-404; ACN-precipitation; HPLC-Fluor (355/410 nm) 0.5 ng/ml
[4] J Chromatogr 423 (1987) 351-357, extraction; HPLC-Fluor (368/415 nm) 0.5 ng/ml

Amoxicillin

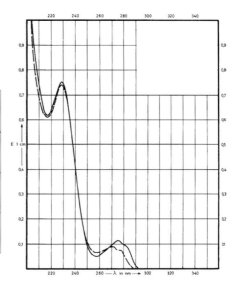

MW: 365.4

solvent symbol	methanol ———	water —·—·—	0.1M HCl ―――――	0.1M NaOH ‥‥‥
absorption maximum	276 nm 229 nm		272 nm 228 nm	
$E_{1cm}^{1\%}$	36 236		27 232	
ε	1390 9160		1050 9000	

C_{max}-value	protein binding (PB)	elimination half life $t_{1/2}$	pKa-value	Ionisation MS	Parent Ion
					Fragment Ion(s)
500 mg oral → 8 µg/ml	20 %	1 h	2.4/7.4/ 9.6	pos	366
					349/208/134/ 114

detection limits	LOD	precipitation	LLE	SPE
UV	1 ng (230 nm)		15 ng/ml [3]	
FL	10 pg [1]	50 ng/ml [1] 100 ng/ml [2]		
ECD	0.1 ng			
MS/MS		120 ng/ml [4]		

(1) J Chromatogr 506 (1990) 417-421; HClO4-precipitation HPLC Fluor (255/400 nm) (derivatization ESA) 50 ng/ml, 10 pg LOD
(2) J Chromatogr A 812 (1998) 221-226; HClO4-precipitation HPLC-Fluor (395/485 nm) (Deriv.) 100 ng/ml
(3) J Pharm Biomed Anal 45 (2007) 531-534, extraction; HPLC-UV (228 nm) 15 ng/ml
(4) J Chromatogr B 813 (2004) 121-127, ACN-precipitation; HPLC-MS 120 ng/ml

Amphetamine

MW: 135.2

no UV
no E1/1

C_{max}-value	protein binding (PB)	elimination half life $t_{1/2}$	pKa-value	Ionisation MS	Parent Ion
					Fragment Ion(s)
10 mg oral → 20 ng/ml after 5 mg Selegelin oral → 1.5 ng/ml	15 – 40 %	4 – 8 h	9.9	pos	136
					91/65/119

detection limits	LOD	precipitation	LLE	SPE
UV	10 ng (255 nm)			15 ng/ml [4]
FL			0.2 ng/ml [1]	
ECD				
MS/MS			0.5 ng/ml [3]	0.3 ng/ml [2]

(1) J Liqu Chrom Rel Technol 20 (1997) 797-809; extraction, HPLC-Fluor (250/395 nm) (nach Deriv.) 0.2 ng/ml
(2) Anal Chim Acta 538 (2005) 49-56, SPE; HPLC-MS/MS 0.3 ng/ml
(3) J Chr B 854 (2007) 48-46, Extraktion; HPLC-MS/MS 0.5 ng/ml
(4) Chromatographia 60 (2004) 537-544, SPE derivatization HPLC-UV (265 nm) 15 ng/ml

Ampicillin

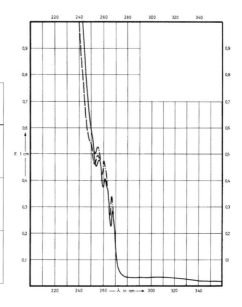

MW: 349.4

solvent symbol	methanol ———	water —·—·—	0.1M HCl — — — —	0.1M NaOH · · · · · ·
absorption maximum	267 nm 261 nm 256 nm	267 nm 261 nm 256 nm	267 nm 261 nm 256 nm	decom- position observed
$E_{1cm}^{1\%}$	5.4 7.6 8.9	6.3 8.9 9.9	6.5 9.0 9.7	
ε	185 265 310	220 310 350	225 315 340	

C_{max}-value	protein binding (PB)	elimination half life $t_{1/2}$	pKa-value	Ionisation MS	Parent Ion Fragment Ion(s)
500 mg oral → 3 µg/ml	20 %	1 – 2 h	2.5/7.3	pos	350 ——— 106/192

detection limits	LOD	precipitation	LLE	SPE
UV	10 ng (240 nm)	5 µg/ml [2]		0.1 µg/ml [1]
FL				
ECD				
MS/MS				

(1) HPLC-UV (220 nm), 0.1 µg/ml
(2) Liqu Chr 8 (1995) 1455-1464, ACN-precipitation; HPLC-UV (220 nm) 5 µg/ml

Bacitracin

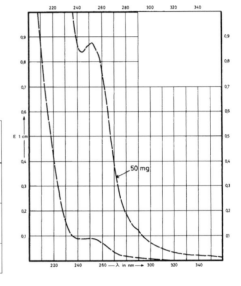

Bacitracin A   MW: 1422.7

solvent symbol	methanol ———	water —·—·—	0.1M HCl —————	0.1M NaOH ······
absorption maximum			252 nm	
$E_{1cm}^{1\%}$			18	
ε				

C_{max}-value	protein binding (PB)	elimination half life $t_{1/2}$	pKa-value	Ionisation MS	Parent Ion ——— Fragment Ion(s)
	40 – 90 %			pos (2++)	712 ——— 119

detection limits	LOD	precipitation	LLE	SPE
UV	20 ng (252 nm)			
FL				
ECD				
MS/MS		0.2 µg/ml [1]		

(1) J Pharm Biomed Anal 43 (2007) 691-700, ACN-precipitation; HPLC-MS/MS 0.2 µg/ml

Bendroflumethiazide

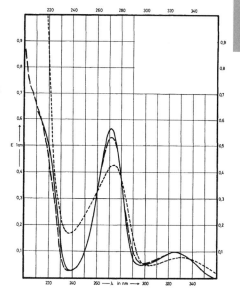

MW: 421.4

solvent symbol	methanol ———	water —·—·—	0.1M HCl —————	0.1M NaOH ······
absorption maximum	324 nm 272 nm		324 nm 273 nm	330 nm 273 nm
$E_{1cm}^{1\%}$	99 560		102 535	87 432
ε	4170 23600		4300 22540	3670 18200

C_{max}-value	protein binding (PB)	elimination half life $t_{1/2}$	pKa-value	Ionisation MS	Parent Ion —————— Fragment Ion(s)
2.5 mg oral	high	3 – 4 h	1.8	pos	422 122/405
				neg	420 328/289/224

detection limits	LOD	precipitation	LLE	SPE
UV	1 ng (272 nm)			
FL		10 ng/ml [3]	1 ng/ml [2]	0.4 ng/ml [1]
ECD				
MS/MS				

(1) SPE, HPLC-Fluor (327/395 nm) 0.4 ng/ml
(2) Chromatographia 63 (2006) 243-248, extraction; HPLC-Fluor 1 ng/ml
(3) Liqu Chromatogr 9 (1986) 2937-2943, ACN-precipitation; HPLC-Fluor 10 ng/ml

Benzbromarone

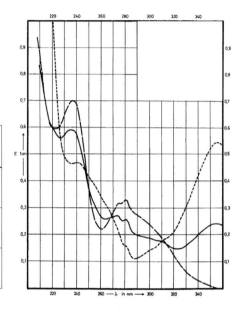

MW: 424.1

solvent symbol	methanol ———	water —·—·—	0.1M HCl —————	0.1M NaOH ······
absorption maximum	355 nm 237 nm		281 nm 237 nm	355 nm 240 nm
$E_{1cm}^{1\%}$	235 562		314 666	513 440
ε	9980 23830		13330 28230	21770 18660

☆ 1M HCl + methanol (1+9)

C_{max}-value	protein binding (PB)	elimination half life $t_{1/2}$	pKa-value	Ionisation MS	Parent Ion Fragment Ion(s)
100 mg oral → 3 µg/ml	high	3 h	6.0	pos	423/425/427

detection limits	LOD	precipitation	LLE	SPE
UV	1 ng (355 nm)		0.1 µg/ml [1]	
FL				
ECD	1 ng			
MS/MS				

(1) Drug Res 31 (1981) 510-512; extraction, HPLC-UV (235 nm) 0.1 µg/ml

Benzoic Acid

MW: 122.1

solvent symbol	methanol ———	water — . — . —	0.1M HCl — — — —	0.1M NaOH . . . . . .
absorption maximum	280 nm 273 nm 227 nm		273 nm 230 nm	269 nm
$E_{1cm}^{1\%}$	61.4 74.3 931		81.0 929	49.7
ε	750 910 11370		990 11340	610

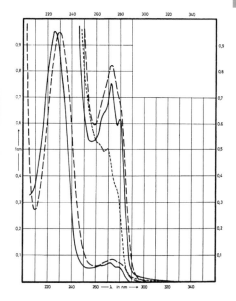

C_{max}-value	protein binding (PB)	elimination half life $t_{1/2}$	pKa-value	Ionisation MS	Parent Ion ——— Fragment Ion(s)
			4.2	pos	123 ——— 61/81/40

detection limits	LOD	precipitation	LLE	SPE
UV	10 ng (273 nm)	50 µg/ml [2]		
FL				
ECD				
MS/MS		1 µg/ml [1]		

(1) HPLC-MS/MS, ACN-precipitation 1 µg/ml
(2) J Chrom 425 (1988) 67-75, ACN-precipitation; HPLC-UV (235 nm) 50 µg/ml

Bezafibrate

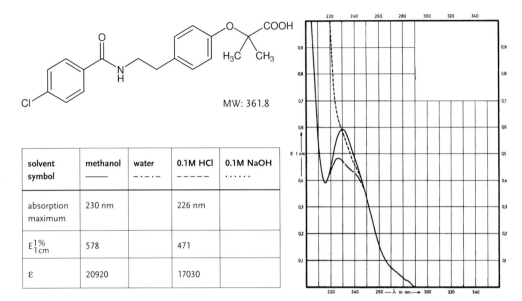

MW: 361.8

solvent symbol	methanol —	water – · – · –	0.1M HCl – – – – –	0.1M NaOH · · · · · ·
absorption maximum	230 nm		226 nm	
$E_{1cm}^{1\%}$	578		471	
ε	20920		17030	

C_{max}-value	protein binding (PB)	elimination half life $t_{1/2}$	pKa-value	Ionisation MS	Parent Ion — Fragment Ion(s)
200 mg oral → 7 µg/ml	95 %	2 h	4.3	pos	362 — 316/121/139
				neg	360 — 274/154

detection limits	LOD	precipitation	LLE	SPE
UV	1 ng (230 nm)	0.1 µg/ml [2]	0.25 µg/ml [3]	
FL				
ECD				
MS/MS				

(2) J Chromatogr Sci 46 (2008) 844-847, HClO₄-precipitation + Meth.; HPLC-UV (235 nm) 0.1 µg/ml
(3) J Chromatogr 344 (1985) 259-265, extraction; HPLC-UV (230 nm) 0.25 µg/ml

Budesonide

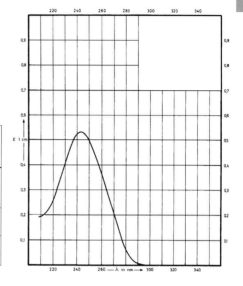

MW: 430.5

solvent symbol	methanol ———	water – · – · –	0.1M HCl – – – –	0.1M NaOH · · · · · ·
absorption maximum	243 nm			
$E_{1cm}^{1\%}$	347			
ε	14950			

C_{max}-value	protein binding (PB)	elimination half life $t_{1/2}$	pKa-value	Ionisation MS	Parent Ion ——— Fragment Ion(s)
250 µg spray → 280 pg/ml	88 %	2.8 h		pos	431 ——— 413/147/173
				neg	489 ——— 357

detection limits	LOD	precipitation	LLE	SPE
UV	1 ng (243 nm)			
FL				
ECD				
MS/MS			5 pg/ml [1]	5 pg/ml [2]/ 6 pg/ml [3]

[1] Extraction, HPLC-MS/MS 5 pg/ml
[2] Anal Chem 79 (2007) 3786-3793, SPE; HPLC-MS/MS 5 pg/ml
[3] J Chrom 823 (1998) 401-409; SPE; HPLC-MS/MS 6 pg/ml

Butorphanol

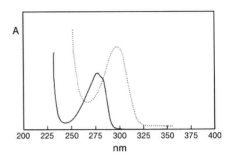

MW: 327.5

no UV
no E1/1

C_{max}-value	protein binding (PB)	elimination half life $t_{1/2}$	pKa-value	Ionisation MS	Parent Ion
					Fragment Ion(s)
15 mg oral → 2 ng/ml	80 %	2 – 4 h	3.7	pos	328
					310

detection limits	LOD	precipitation	LLE	SPE
UV	10 ng (287 nm)			
FL				
ECD	0.1 ng			
MS/MS			0.014 ng/ml [2]	0.2 ng/ml [1]

(1) HPLC-MS/MS 0.2 ng/ml SPE
(2) J Chromatogr B 775 (2002) 57-62, extraction; 14 pg/ml HPLC-MS/MS

Caffeine

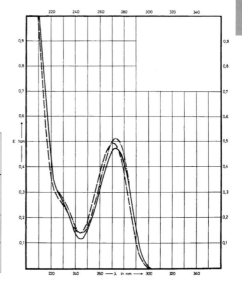

MW: 194.2

solvent symbol	methanol ———	water – · – · –	0.1M HCl – – – –	0.1M NaOH · · · · · ·
absorption maximum	273 nm	273 nm	270 nm	273 nm
$E_{1cm}^{1\%}$	475	515	495	510
ε	9220	10000	9610	9900

C_{max}-value	protein binding (PB)	elimination half life $t_{1/2}$	pKa-value	Ionisation MS	Parent Ion
					Fragment Ion(s)
100 mg oral → 3 µg/ml (usual 2 – 10 µg/ml)	35 %	2 – 10 h	14.0	pos	195
					138/110/83

detection limits	LOD	precipitation	LLE	SPE
UV	1 ng (273 nm)			
FL				
ECD				
MS/MS	2 pg		10 ng/ml [2]	

(2) Chromatographia 67 (2008) 281-285; extraction HPLC-MS 10 ng/ml

Canrenone

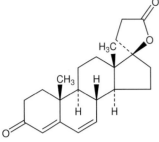

MW: 340.5

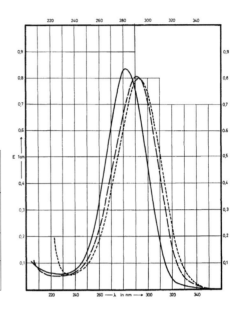

solvent symbol	methanol ———	water —·—·—	0.1M HCl —————	0.1M NaOH ······
absorption maximum	283 nm		292 nm	293 nm
$E_{1cm}^{1\%}$	798		770	768
ε	27130		26210	26140

C_{max}-value	protein binding (PB)	elimination half life $t_{1/2}$	pKa-value	Ionisation MS	Parent Ion ——— Fragment Ion(s)
after 100 mg oral Spironolacton → 350 ng/ml Canrenone (usual 50 – 250 ng/ml	>90 %	15 – 21 h		pos	341 ——— 205/323
				neg	357 ——— 313/339/145

detection limits	LOD	precipitation	LLE	SPE
UV	1 ng (283 nm)		7 ng/ml [1]	6 ng/ml [3]
FL				
ECD			2 ng/ml [2]	
MS/MS				

(1) HPLC-UV (283 nm) 7 ng/ml extraction
(2) J Mass Spectrom 41 (2006) 477-486, nach extraction; HPLC-MS APCI 2 ng/ml
(3) J Chromatogr 574 (1992) 57-64, SPE; HPLC-UV (280 nm) 6 ng/ml

Captopril

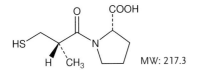

MW: 217.3

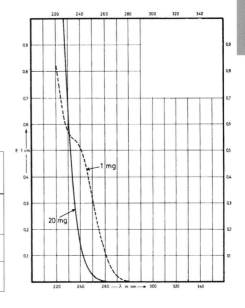

solvent symbol	methanol ———	water — · — · —	0.1M HCl — — — —	0.1M NaOH · · · · · ·
absorption maximum				
$E_{1cm}^{1\%}$				
ε				

C_{max}-value	protein binding (PB)	elimination half life $t_{1/2}$	pKa-value	Ionisation MS	Parent Ion ——— Fragment Ion(s)
100 mg oral → 0.8 µg/ml	30 %	1 – 2 h	3.7/9.8	pos	218 ——— 116/70/75
				neg	216 ——— 182/114

detection limits	LOD	precipitation	LLE	SPE
UV	10 ng (230 nm)			7 ng/ml [2]
FL		2 ng/ml [1]		
ECD				
MS/MS			25 ng/ml [3]	

[1] Precipitation, derivatization HPLC-fluorescence (325/475 nm) 2 ng/ml
[2] J Pharm Biomed Anal 41 (2006) 644-648, derivatization; HPLC-UV (263 nm) 7 ng/ml
[3] J Pharm Biomed Anal 37 (2005) 1073-1080, extraction; HPLC-MS/MS 25 ng/ml

Carbamazepine

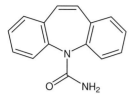

MW: 236.3

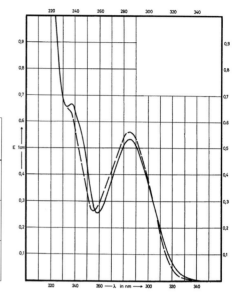

solvent symbol	methanol ———	water —·—·—	0.1M HCl —————	0.1M NaOH ······
absorption maximum	285 nm 237 nm		285 nm	285 nm
$E_{1cm}^{1\%}$	454 568		480	480
ε	10740 13410		11330	11330

C_{max}-value	protein binding (PB)	elimination half life $t_{1/2}$	pKa-value	Ionisation MS	Parent Ion ——— Fragment Ion(s)
200 mg oral → 4 µg/ml (usual 3 – 12 µg/ml)	75 %	10 – 30 h		pos	237 ——— 194/192/179

detection limits	LOD	precipitation	LLE	SPE
UV	1 ng (285 nm)	0.05 µg/ml [1]	0.01 µg/ml [2]	
FL				
ECD				
MS/MS				

[1] HPLC-UV (305 nm) 0.05 µg/ml ACN-precipitation
[2] Anal Bioanal Chem 386 (2006) 1931-1936, extraction; HPLC-UV (210 nm) 0.01 µg/ml

Carvedilol

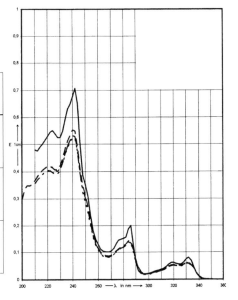

MW: 406.5

solvent symbol	methanol ———	water — · — · —	0.1M HCl — — — —	0.1M NaOH · · · · · ·
absorption maximum	332 nm 286 nm 242 nm	332 nm 285 nm 241 nm	332 nm 285 nm 241 nm	332 nm 285 nm 241 nm
$E_{1cm}^{1\%}$	163 401 1415	116 276 1030	121 287 1085	123 290 1065
ε	6600 16300 57500	4700 11200 41800	4900 11700 44100	5000 11800 43300

C_{max}-value	protein binding (PB)	elimination half life $t_{1/2}$	pKa-value	Ionisation MS	Parent Ion ——————— Fragment Ion(s)
20 mg oral → 50 ng/ml	98 %	4 – 8 h	3.0	pos	407 ——————— 222/100/194

detection limits	LOD	precipitation	LLE	SPE
UV	1 ng (286 nm)		5 ng/ml [3]	
FL	0.1 ng [1]		2 ng/ml [1]/ 1 ng/ml [2]	
ECD				
MS/MS			0.1 ng/ml [4]	

(1) HPLC-Fluor (285/340 nm) 2 ng/ml extraction
(2) J Chrom 857 (2007) 219-223; HPLC-Fluor (282/340 nm) after extraction 1 ng/ml
(3) J Liqu Chrom Relat Technol 30 (2007) 1677-1685; HPLC-UV (238 nm) after extr. 5 ng/ml
(4) J Chrom 822 (2005) 253-262; HPLC-MS/MS after extraction 0.1 ng/ml

Chlorthalidone (Blood)

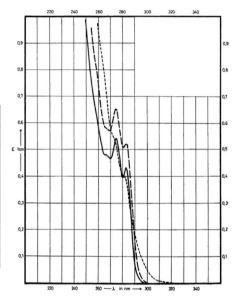

MW: 338.8

solvent symbol	methanol ———	water –·–·–	0.1M HCl – – – –	0.1M NaOH ......
absorption maximum	283 nm 274 nm		283 nm 275 nm	
$E_{1cm}^{1\%}$	43 54		52 65	
ε	1460 1830		1760 2200	

C_{max}-value	protein binding (PB)	elimination half life $t_{1/2}$	pKa-value	Ionisation MS	Parent Ion —————— Fragment Ion(s)
25 mg oral → 1.5 µg/ml whole blood (ca. 98 % erythrocytes)	75 % plasma	35 – 70 h	9.4	pos	339 —————— 322/243

detection limits	LOD	precipitation	LLE	SPE
UV	10 ng (274 nm)	25 ng/ml [3]	40 ng/ml [1]/ 10 ng/ml [2]	
FL				
ECD				
MS/MS				

(1) HPLC-UV (230 nm) 40 ng/ml extraction
(2) J Chromatogr 698 (1997) 187-194 plasma, extraction; HPLC-UV (225 nm) 10 ng/ml
(3) J Chromatogr 416 (1987) 420-425 whole blood ; HClO4 precipitation; HPLC-UV (214 nm) 25 ng/ml

Ciclesonide

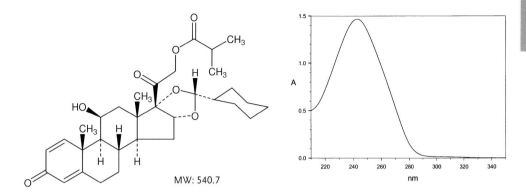

MW: 540.7

no UV
no E1/1

C$_{max}$-value	protein bin-ding (PB)	elimination half life t$_{1/2}$	pKa-value	Ionisation MS	Parent Ion
					Fragment Ion(s)
after inhalation >100 pg/ml	>99 %	0.7 h (3.5 h Des-CIC)		neg	599 (acetic adduct) 339

detection limits	LOD	precipitation	LLE	SPE
UV	1 ng (237 nm)			
FL				
ECD				
MS/MS			10 pg/ml [1]	

(1) J Chrom 869 (2008), 84-92, HPLC-MS/MS after extraction 10 pg/ml

Cimetidine

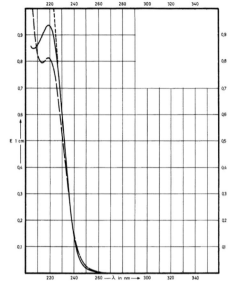

MW: 252.3

solvent symbol	methanol ——	water —·—·—	0.1M HCl -----	0.1M NaOH ......
absorption maximum	219 nm		219 nm	
$E_{1cm}^{1\%}$	810		790	
ε	20440		19930	

C_{max}-value	protein binding (PB)	elimination half life $t_{1/2}$	pKa-value	Ionisation MS	Parent Ion ———— Fragment Ion(s)
400 mg oral → 2 µg/ml (usual 0.5 – 1 µg/ml)	13 – 26 %	1 – 3 h	6.8	pos	253 ———— 159/95/117
				neg	251 ———— 97/123

detection limits	LOD	precipitation	LLE	SPE
UV	1 ng (219 nm)		2 ng/ml [3]	50 ng/ml [1]
FL				
ECD				
MS/MS				5 ng/ml [2]

[1] J Chromatogr 273 (1983) 449-452, SPE; HPLC-UV (220 nm) 50 ng/ml
[2] Biomed Chromatogr 13 (1999) 455-461; HPLC-MS/MS 5 ng/ml
[3] J Chromatogr 227 (1982) 521-525, extraction; HPLC-UV (250 nm) 2 ng/ml

Cinnarizine

MW: 368.5

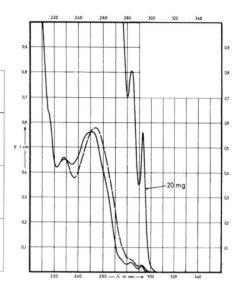

-20 mg

solvent symbol	methanol ———	water –·–·–	0.1M HCl – – – –	0.1M NaOH ......
absorption maximum	293 nm 284 nm 251 nm		254 nm 228 nm	
$E_{1cm}^{1\%}$	28 40 562		578 458	
ε	1030 1480 20700		21300 16880	

C_{max}-value	protein binding (PB)	elimination half life $t_{1/2}$	pKa-value	Ionisation MS	Parent Ion —————— Fragment Ion(s)
25 mg oral → 100 ng/ml		5 h	5.4	pos	369 —————— 167/152

detection limits	LOD	precipitation	LLE	SPE
UV	1 ng (250 nm)		2 ng/ml [1]	
FL			1 ng/ml [3]	
ECD				
MS/MS		2 ng/ml [2]		

[1] J Chromatogr 227 (1982) 521-525, extraction; HPLC-UV (250 nm) 2 ng/ml
[2] Chromatographia 67 (2008) 583-590, precipitation; HPLC-MS/MS 2 ng/ml
[3] Chromatographia 36 (1993) 356-358, extraction; HPLC-Fluor (245/310 nm) 1 ng/ml

Clarithromycin

MW: 748.0

no UV
no E1/1

C$_{max}$-value	protein binding (PB)	elimination half life t$_{1/2}$	pKa-value	Ionisation MS	Parent Ion
					Fragment Ion(s)
250 mg oral → 900 ng/ml	42 – 50 %	3 – 4 h	3.2	pos	748
					590/158

detection limits	LOD	precipitation	LLE	SPE
UV	100 ng (290 nm)		31 ng/ml [3]	
FL				
ECD	1 ng			100 ng/ml [4]
MS/MS		10 ng/ml [2]	20 ng/ml [1]	

[1] HPLC-MS 20 ng/ml extraction
[2] J Pharm Biomed Anal 43 (2007) 1460-1464, ACN-precipitation; HPLC-MS/MS 10 ng/ml
[3] J Chromatogr 817 (2005) 193-197, extraction; HPLC-UV (205 nm) 31 ng/ml
[4] Talanta 54 (2001), 377-382 SPE; HPLC-ECD 100 ng/ml

Codeine

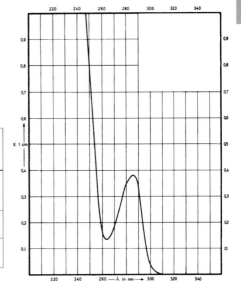

MW: 299.4

solvent symbol	methanol ———	water –·–·–	0.1M HCl – – – –	0.1M NaOH · · · · · ·
absorption maximum	286 nm			
$E_{1cm}^{1\%}$	52.8			
ε	1580			

C_{max}-value	protein binding (PB)	elimination half life $t_{1/2}$	pKa-value	Ionisation MS	Parent Ion ——— Fragment Ion(s)
50 mg oral → 0.2 µg/ml (usual 0.025 – 0.15 µg/ml)	7 – 25 %	2 – 4 h	8.2	pos	300 ——— 165/153

detection limits	LOD	precipitation	LLE	SPE
UV	1 ng (286 nm)			10 ng/ml [1]
FL			10 ng/ml [2]	
ECD				5 ng/ml [3]
MS/MS				

[1] J Pharm Sci 73 (1984) 1556-1558, SPE; HPLC-UV (220 nm) 10 ng/ml
[2] J Liqu Chromatogr 16 (1993) 2325-2334, extraction; HPLC-fluorescence (285/345 nm) 10 ng/ml
[3] J Chromatogr 570 (1991) 309-320, SPE; HPLC-ECD 5 ng/ml

Dextrorphan

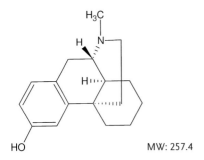

MW: 257.4

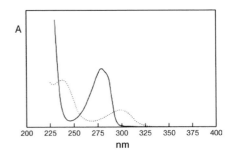

no E1/1

C_{max}-value	protein binding (PB)	elimination half life $t_{1/2}$	pKa-value	Ionisation MS	Parent Ion
					Fragment Ion(s)
after 60 mg Dextrometorphan → 700 ng/ml		3 h	3.4	pos	258
					199

detection limits	LOD	precipitation	LLE	SPE
UV	10 ng (280 nm)			
FL			0.3 ng/ml [2]	6 ng/ml [1]
ECD	1 ng			
MS/MS			0.01 ng/ml [3]	

(1) J Chromatogr 420 (1987) 217-222, SPE; HPLC-Fluor. (275/305 nm) 6 ng/ml
(2) J Chrom 859 (2007) 141-146, extraction; HPLC-Fluor 0.3 ng/ml
(3) J Pharm Biomed Anal 43 (2007) 586-600, extraction; HPLC-MS/MS 0.01 ng/ml

Diazepam

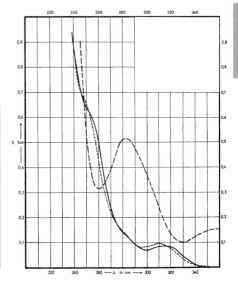

MW: 284.7

solvent symbol	methanol ———	water —·—·—	0.1M HCl – – – –	0.1M NaOH ······
absorption maximum	315 nm		360 nm 282 nm	310 nm
$E_{1cm}^{1\%}$	79		142 477	84
ε	2250		4040 13580	2390

C_{max}-value	protein binding (PB)	elimination half life $t_{1/2}$	pKa-value	Ionisation MS	Parent Ion ——— Fragment Ion(s)
5 mg oral → 100 ng/ml (usual 200 – 500 ng/ml)	98 – 99 %	20 – 100 h	3.3	pos	285 ——— 193/154/222

detection limits	LOD	precipitation	LLE	SPE
UV	1 ng (230 nm)			
FL				
ECD				
MS/MS		1 ng/ml [2]		

(2) J Chromatogr B 874 (2008) 42-50, ACN-precipitation; HPLC-MS/MS 1 ng/ml

D
–
F

Diclofenac

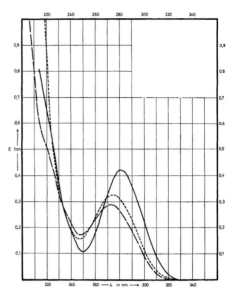

MW: 296.2

solvent symbol	methanol ——	water —·—·—	0.1M HCl – – – –	0.1M NaOH ······
absorption maximum	282 nm		274 nm	275 nm
$E_{1cm}^{1\%}$	425		288	327
ε	12030		8150	9250

C_{max}-value	protein binding (PB)	elimination half life $t_{1/2}$	pKa-value	Ionisation MS	Parent Ion ——— Fragment Ion(s)
50 mg oral → 800 ng/ml (usual 50 – 2500 ng/ml)	>99 %	1 – 2 h	3.1	pos	296 ——— 214/250/178

detection limits	LOD	precipitation	LLE	SPE
UV	1 ng (282 nm)		10 ng/ml [1a]	
FL				
ECD			5 ng/ml [3]	
MS/MS	10 pg		1 ng/ml [1b]	10 ng/ml [2]

(1a) Drug Design and Delivery 4 (1989) 303-311, extraction; HPLC-UV (300 nm) 10 ng/ml
(1b) (Int J Clin Pharm Ther 42 (2004) 353-359, extraction; HPLC-MS/MS 1 ng/ml)
(2) Anal Bioanal Chem 384 (2006) 1501-1505, SPE; HPLC-MS 10 ng/ml
(3) Biomed Chromatogr. 20 (2006) 119-124, extraction; HPLC-ECD 5 ng/ml

Dihydralazine

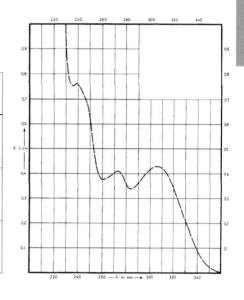

MW: 190.2

solvent symbol	methanol —	water – · – · –	0.1M HCl – – – – –	0.1M NaOH · · · · · ·
absorption maximum			306 nm 274 nm 240 nm	
$E_{1cm}^{1\%}$ *			213 204 377	
ε			6130 5870 10870	

* Referred to dried

C_{max}-value	protein binding (PB)	elimination half life $t_{1/2}$	pKa-value	Ionisation MS	Parent Ion ——— Fragment Ion(s)
25 mg oral → 50 ng/ml		1 – 2 h	0.2	pos	191 ——— 129/102/174

detection limits	LOD	precipitation	LLE	SPE
UV	1 ng (306 nm)			
FL				
ECD				
MS/MS		0.5 ng/ml [1]		

(1) J Pharm Biomed Anal 43 (2007) 631-645; TCA-precipitation; HPLC-MS/MS 0.5 ng/ml

Diphenhydramine

MW: 255.4

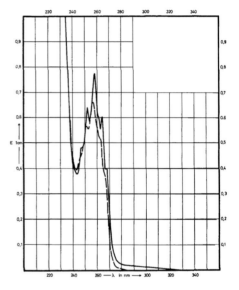

solvent symbol	methanol ———	water —·—·—	0.1M HCl —————	0.1M NaOH ······
absorption maximum	264 nm 258 nm 252 nm		258 nm 252 nm	
$E_{1cm}^{1\%}$	12.1 15.5 12.8		13.3 11.4	
ε	353 452 374		388 333	

C_{max}-value	protein binding (PB)	elimination half life $t_{1/2}$	pKa-value	Ionisation MS	Parent Ion ——— Fragment Ion(s)
50 mg oral → 50 ng/ml (usual 80 – 400 ng/ml)	78 %	2 – 9 h	3.1	pos	256 ——— 165/152/167

detection limits	LOD	precipitation	LLE	SPE
UV	10 ng (258 nm)			
FL			0.5 ng/ml [3]	
ECD				
MS/MS			1 ng/ml [2]	

(2) J Chromatogr B 854 (2007) 48-56, extraction; HPLC-MS/MS 1 ng/ml
(3) Pharm Res 8 (1991) 1448-1451, extraction; HPLC-fluorescence (230/560 nm) 0.5 ng/ml

Dipyridamole

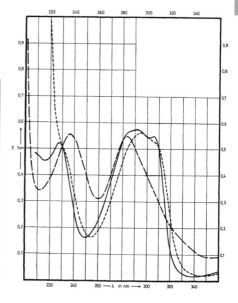

MW: 504.6

solvent symbol	methanol ——	water —·—·—	0.1M HCl – – – –	0.1M NaOH ······
absorption maximum	408 nm 292 nm 228 nm		400 nm 283 nm 237 nm	414 nm 295 nm
$E_{1cm}^{1\%}$	165 570 520		134 546 554	155 558
ε	8330 28760 26190		6760 27550 27950	7820 28160

C_{max}-value	protein binding (PB)	elimination half life $t_{1/2}$	pKa-value	Ionisation MS	Parent Ion ——— Fragment Ion(s)
100 mg oral → 2 µg/ml	>90 %	12 h	6.4	pos	505 ——— 429/385

detection limits	LOD	precipitation	LLE	SPE
UV	1 ng (300 nm)		30 ng/ml [1]	
FL				
ECD				0.5 ng/ml [3]
MS/MS			10 ng/ml [2]	

[1] HPLC-UV (290 nm) 30 ng/ml extraction
[2] Biomed Chromatogr 22 (2008) 149-156, extraction, HPLC-MS 10 ng/ml
[3] J Liqu Chromatogr 17 (1994) 1837-1848, SPE, HPLC-ECD 0.5 ng/ml

Doxorubicin

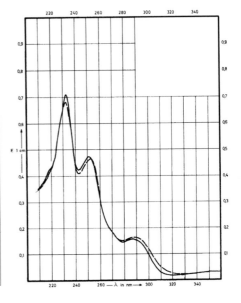

MW: 543.5

solvent symbol	methanol ———	water —·—·—	0.1M HCl – – – –	0.1M NaOH ······
absorption maximum	476 nm 251 nm 233 nm		478 nm 253 nm 232 nm	decomposition observed
$E_{1cm}^{1\%}$	215 441 660		205 429 634	
ε	12500 25600 38800		11900 24900 36800	

C_{max}-value	protein binding (PB)	elimination half life $t_{1/2}$	pKa-value	Ionisation MS	Parent Ion ——— Fragment Ion(s)
1.5 mg/kg iv	76 %	3 and 40 h	8.2/10.2	pos	544 ——— 361/321/397
				neg	542 ——— 395

detection limits	LOD	precipitation	LLE	SPE
UV	1 ng (290 nm)			
FL		5 ng/ml [1]/ 5 ng/ml [2]		
ECD				
MS/MS		20 ng/ml [3]		

(1) HPLC-Fluor. precipitation 5ng/ml
(2) Biomed Chromatogr 22 (2008) 1252-1258, ACN-precipitation; HPLC-Fluor (475/580 nm) 5 ng/ml
(3) Talanta 74 (2008) 887-895, Aceton-precipitation; HPLC-MS/MS 20 ng/ml

Doxycycline

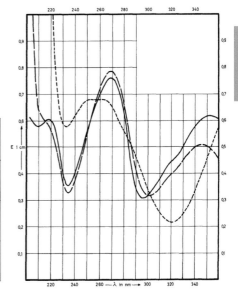

MW: 462.5

solvent symbol	methanol ——	water –·–·–	0.1M HCl – – – –	0.1M NaOH ······
absorption maximum	352 nm 269 nm		346 nm 269 nm	376 nm 255 nm
$E_{1cm}^{1\%}$	297 368		246 381	329 326
ε	14290 17700		11850 18330	15820 15680

C_{max}-value	protein binding (PB)	elimination half life $t_{1/2}$	pKa-value	Ionisation MS	Parent Ion ———— Fragment Ion(s)
100 mg oral → 3 µg/ml (usual 3 – 5.3 µg/ml)	82 – 90 %	22 h	3.5/7.7/ 9.5	pos	446 (decomp.) ———— 156/428

detection limits	LOD	precipitation	LLE	SPE
UV	1 ng (352 nm)	0.5 µg/ml [1]/ 0.4 µg/ml [2]		
FL				
ECD				
MS/MS				

[1] HPLC-UV (370 nm) 0.5 µg/ml TCA-precipitation
[2] J Chromatogr 1031 (2004) 295-301, ACN-precipitation; HPLC-UV (347 nm) 0.4 µg/ml

Duramycin (Moli 1901)

no structure
MW: 2013.6
no UV
no E1/1

C_{max}-value	protein binding (PB)	elimination half life $t_{1/2}$	pKa-value	Ionisation MS	Parent Ion
					Fragment Ion(s)
>1 – 10 ng/ml				pos (2^{++})	1007
					1007

detection limits	LOD	precipitation	LLE	SPE
UV	100 ng			
FL				
ECD				
MS/MS	25 pg [1]	1 ng/ml [1]		

(1) Precipitation; HPLC-MS/MS 1 ng/ml

Erythromycin

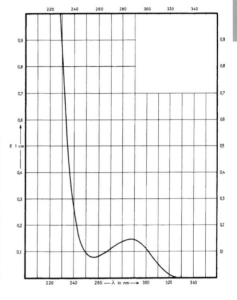

MW: 733.9   Erythromycin A

solvent symbol	methanol ————	water —·—·—	0.1M HCl ————	0.1M NaOH ······
absorption maximum				
$E_{1cm}^{1\%}$				
ε				

C_{max}-value	protein binding (PB)	elimination half life $t_{1/2}$	pKa-value	Ionisation MS	Parent Ion ———— Fragment Ion(s)
500 mg oral → 2 µg/ml	70 – 80 %	1 – 3 h	8.9	pos	734 ———— 576/158/83

detection limits	LOD	precipitation	LLE	SPE
UV	100 ng (290 nm)			
FL				
ECD	1 ng		100 ng/ml [3]	
MS/MS		2 ng/ml [2]	10 ng/ml [1]	

(1) Extraction; HPLC-MS 10 ng/ml
(2) J Chromatogr B 817 (2005) 153-158, ACN-precipitation, HPLC-MS 2 ng/ml
(3) J Chromatogr B 738 (2000) 405-411, extraction; HPLC-ECD 100 ng/ml

Fluticasone Propionate

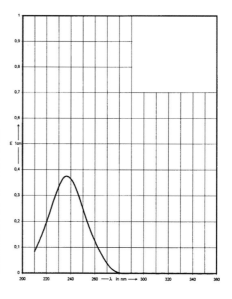

MW: 500.6

solvent symbol	methanol ———	water – · – · –	0.1M HCl – – – –	0.1M NaOH · · · · · ·
absorption maximum	237 nm			
$E_{1cm}^{1\%}$	372			
ε	18600			

C_{max}-value	protein binding (PB)	elimination half life $t_{1/2}$	pKa-value	Ionisation MS	Parent Ion —————— Fragment Ion(s)
>50 – 100 pg/ml after inhalation	81 – 95 %	3.5 h		pos	501 —————— 313/293/205

detection limits	LOD	precipitation	LLE	SPE
UV	1 ng (237 nm)			
FL				
ECD				
MS/MS	1 pg [1]		10 pg/ml [1]	20 pg/ml [2]

(1) J Chromatogr B 869(2008) 84-92, extraction; HPLC-MS/MS 10 pg/ml
(2) J Chromatogr B 761 (2001) 177-185, SPE; HPLC-MS/MS 20 pg/ml

Formoterol

OH

R,R-Form

HO

HN–CH
‖
O

MW: 344.4

A

200 225 250 275 300 325 350 375 400
nm

no UV
no E1/1

C_{max}-value	protein binding (PB)	elimination half life $t_{1/2}$	pKa-value	Ionisation MS	Parent Ion Fragment Ion(s)
after inhalation of 36 µg → 30 pg/ml		8 h	1.4	pos	345 149

detection limits	LOD	precipitation	LLE	SPE
UV	1 ng			
FL				
ECD				20 pg/ml [2]
MS/MS				0.4 pg/ml [1]

(1) J Chromatogr B 830 (2006) 25-34, SPE; HPLC-MS/MS 0.4 pg/ml
(2) Ther Drug Monit 16 (1994) 196-199, SPE; HPLC-ECD 20 pg/ml

D – F

Furosemide

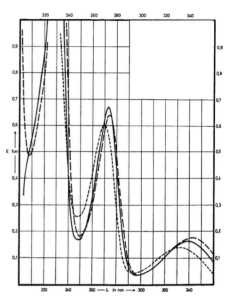

MW: 330.7

solvent symbol	methanol ──	water ─·─·─	0.1M HCl ─ ─ ─ ─	0.1M NaOH ······
absorption maximum	337 nm 273 nm 233 nm		340 nm 274 nm 235 nm	335 nm 270 nm
$E_{1cm}^{1\%}$	159 671 1390		169 638 1430	131 586
ε	5260 22190 45970		5590 21100 47290	4330 19380

C_{max}-value	protein binding (PB)	elimination half life $t_{1/2}$	pKa-value	Ionisation MS	Parent Ion / Fragment Ion(s)
80 mg oral → 2.3 µg/ml (usual 2 – 5 µg/ml)	97 %	1 – 3 h	3.9	pos	331 / 119/303/299
				neg	329 / 285/205/78

detection limits	LOD	precipitation	LLE	SPE
UV	1 ng (273 nm)			100 ng/ml [3]
FL		27 ng/ml [2]		
ECD				
MS/MS				

(2) Biomed Chromatogr 21 (2007) 40-47, ACN-precipitation; HPLC-Fluor (235/402 nm) 27 ng/ml
(3) J Pharm Biomed Anal 33 (2003) 699-709; HPLC-UV (235 nm) 100 ng/ml

Glibenclamide

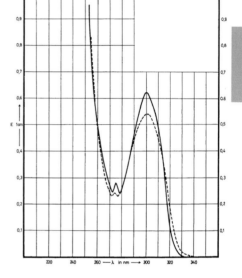

MW: 494.0

solvent symbol	methanol ———	water —·—·—	0.1M HCl —————	0.1M NaOH ······
absorption maximum	300 nm 274 nm			301 nm 274 nm
$E_{1cm}^{1\%}$	62.4 28.2			54.3 24.1
ε	3080 1390			2680 1190

C_{max}-value	protein binding (PB)	elimination half life $t_{1/2}$	pKa-value	Ionisation MS	Parent Ion Fragment Ion(s)
3.5 mg oral → 150 ng/ml (usual maximum 100 – 300 ng/ml)	99 %	5 – 16 h	5.3	pos	494 369/169/304

detection limits	LOD	precipitation	LLE	SPE
UV	10 ng (300 nm)	300 ng/ml [2]		5 ng/ml [3]
FL	0.1 ng/ml [1]		4 ng/ml [1]	
ECD				
MS/MS				

[1] Clin Drug Invest 15 (1998) 253-260; HPLC-Fluor (235/390 nm) 4 ng/ml
[2] Biomed Chromatogr 20 (2006) 1043-1048, ACN-precipitation; HPLC-UV (260 nm) 300 ng/ml
[3] J Chromatogr B 682 (1996) 364-370, SPE, HPLC-UV (230 nm) 5 ng/ml

G – K

Hydrochlorothiazide

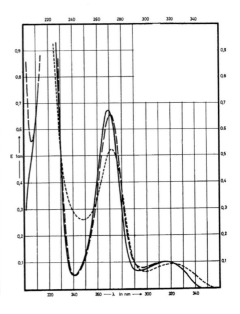

MW: 297.7

solvent symbol	methanol ———	water —·—·—	0.1M HCl – – – –	0.1M NaOH · · · · · ·
absorption maximum	315 nm 269 nm		315 nm 271 nm	320 nm 272 nm
$E_{1cm}^{1\%}$	99 668		99 651	91 523
ε	2950 19890		2950 19390	2710 15570

C_{max}-value	protein binding (PB)	elimination half life $t_{1/2}$	pKa-value	Ionisation MS	Parent Ion
					Fragment Ion(s)
50 mg oral → 300 ng/ml (usual 50 – 160 ng/ml)	60 %	2.5 – 10 h	7.0/9.2	neg	296
					205/269/78

detection limits	LOD	precipitation	LLE	SPE
UV	1 ng (269 nm)		5 ng/ml [1]/ 10 ng/ml [3]	
FL				
ECD				
MS/MS		1 ng/ml [2]		

[1] Extraction; HPLC-UV (315 nm) 5 ng/ml
[2] J Chromatogr B 852 (2007) 436-442, ACN-precipitation; HPLC-MS/MS 1 ng/ml
[3] J Pharm Biomed Anal 41 (2006) 644-648, extraction; HPLC-UV (260 nm) 10 ng/ml

Ibuprofen

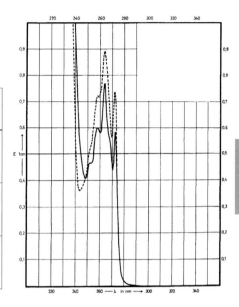

MW: 206.3

solvent symbol	methanol ———	water —·—·—	0.1M HCl ————	0.1M NaOH ......
absorption maximum	272 nm 264 nm 258 nm			272 nm 264 nm 258 nm
$E_{1cm}^{1\%}$	11.2 14.5 11.3			15.4 18.4 15.0
ε	230 300 233			320 380 310

C_{max}-value	protein binding (PB)	elimination half life $t_{1/2}$	pKa-value	Ionisation MS	Parent Ion ——— Fragment Ion(s)
200 mg oral → 20 µg/ml (usual 15 – 30 µg/ml)	99 %	2 h	3.8	neg	206 ——— 161/91/119

detection limits	LOD	precipitation	LLE	SPE
UV	10 ng (272 nm)			
FL			0.1 µg/ml [1a] 0.12 µg/ml [1b]	
ECD				
MS/MS			0.12 µg/ml [2]	

[1a] Extraction; HPLC-fluorescence (225/290 nm) 0.1 µg/ml
[1b] Eur J Clin Pharmacol 48 (1995) 505-511, extraction; HPLC-fluorescence (225/290 nm) 0.2 µg/ml enantiomer separation
[2] J Chromatogr B 796 (2003) 413-420, extraction; HPLC-MS/MS 0.12 µg/ml enantiomer separation

G – K

Indomethazin

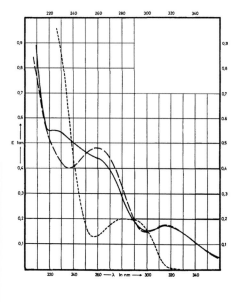

MW: 357.8

solvent symbol	methanol ———	water –·–·–	0.1M HCl – – – –	0.1M NaOH ·····
absorption maximum	316 nm		316 nm 260 nm	280 nm
$E_{1cm}^{1\%}$	172		177 487	201
ε	6150		6330 17420	7190

C_{max}-value	protein binding (PB)	elimination half life $t_{1/2}$	pKa-value	Ionisation MS	Parent Ion / Fragment Ion(s)
50 mg oral → 1.5 µg/ml (usual 0.8 – 2.5 µg/ml)	90 – 99 %	3 – 15 h	4.5	pos	358 / 139

detection limits	LOD	precipitation	LLE	SPE
UV	1 ng (316 nm)	25 ng/ml [2]	50 ng/ml [1]	
FL				
ECD				
MS/MS				

[1] HPLC-UV (320 nm) extraction 50 ng/ml
[2] J Chromatogr B 830 (2006) 364-367, ACN-precipitation; HPLC-UV (270 nm) 25 ng/ml

Isosorbid-5-mononitrate

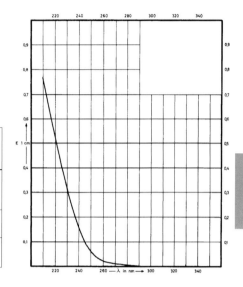

MW: 191.1

solvent symbol	methanol ———	water – · – · –	0.1M HCl – – – –	0.1M NaOH · · · · · ·
absorption maximum				
$E_{1cm}^{1\%}$				
ε				

C_{max}-value	protein binding (PB)	elimination half life $t_{1/2}$	pKa-value	Ionisation MS	Parent Ion / Fragment Ion(s)
20 mg oral → 600 ng/ml	>5 %	3 – 7 h		pos	192 ———

detection limits	LOD	precipitation	LLE	SPE
UV	1 ng (220 nm)		5 ng/ml [1]	
FL				
ECD				
MS/MS			1 ng/ml [2]	

(1) HPLC-UV (210 nm) 5 ng/ml extraction
(2) J Chromatogr B 846 (2007) 323-328, extraction; HPLC-MS/MS 1 ng/ml

Itraconazole

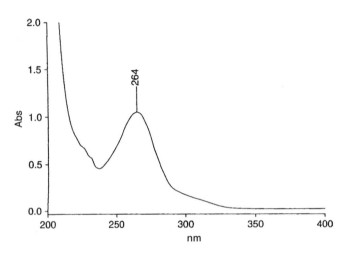

MW: 705.6

no UV
no E1/1

C$_{max}$-value	protein binding (PB)	elimination half life t$_{1/2}$	pKa-value	Ionisation MS	Parent Ion
					Fragment Ion(s)
200 mg daily → 2 µg/ml (usual 0.2 – 2 µg/ml)	99.8 %	20 h	6.2	pos	705
					392/432/450

detection limits	LOD	precipitation	LLE	SPE
UV	1 ng		3 ng/ml [2]	
FL		50 ng/ml [1]		
ECD				
MS/MS				

[1] Biomed Chromatogr 20 (2006) 343-348, ACN-precipitation; HPLC-Fluor (250/380 nm) 50 ng/ml
[2] Ther Drug Monit 28 (2006) 526-531, extraction; HPLC-UV (263 nm) 3 ng/ml

Ketoconazole

MW: 531.4

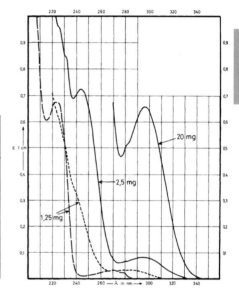

solvent symbol	methanol ———	water –·–·–	0.1M HCl – – – –	0.1M NaOH ......
absorption maximum	296 nm 244 nm		270 nm 223 nm	287 nm
$E^{1\%}_{1cm}$	32 280		27 530	29
ε	1700 14900		1420 28000	1500

C_{max}-value	protein binding (PB)	elimination half life $t_{1/2}$	pKa-value	Ionisation MS	Parent Ion
					Fragment Ion(s)
400 mg oral → 6.5 µg/ml	99 %	6 – 10 h	4.5	pos	531
					489/148/219

detection limits	LOD	precipitation	LLE	SPE
UV	1 ng (244 nm)			15 ng/ml [2]
FL				
ECD				
MS/MS			20 ng/ml [1] / 20 ng/ml [3]	

(1) HPLC-MS/MS extraction 20 ng/ml
(2) J Chromatog. B 839 (2006) 62-67; HPLC-UV (240 nm) 15 ng/ml
(3) J Chromatogr B 774 (2002) 67-78, extraction; HPLC-MS/MS 20 ng/ml

G
–
K

Lansoprazole

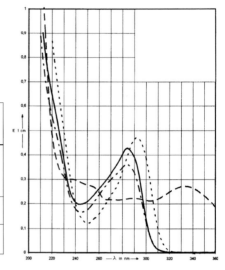

MW: 369.4

solvent symbol	methanol —	water — · — · —	0.1M HCl — — — —	0.1M NaOH · · · · · ·
absorption maximum	decom-position observed	decom-position observed	decom-position observed	292 nm
$E_{1cm}^{1\%}$				466
ε				17200

C_{max}-value	protein binding (PB)	elimination half life $t_{1/2}$	pKa-value	Ionisation MS	Parent Ion — Fragment Ion(s)
30 mg oral → 900 µg/ml	97 %	1 – 2 h	2.8	neg	368 — 164

detection limits	LOD	precipitation	LLE	SPE
UV	1 ng (283 nm)		3 ng/ml [3]	15 ng/ml [1]
FL				
ECD				
MS/MS				0.25 ng/ml [2]

(1) SPE, HPLC-UV (285 nm) 15 ng/ml
(2) J Chromatogr B 870 (2008) 38-45; HPLC-MS/MS 0.25 ng/ml
(3) J Chromatogr B 816 (2005) 309-314, extraction; HPLC-UV (285 nm) 3 ng/ml

Levodopa

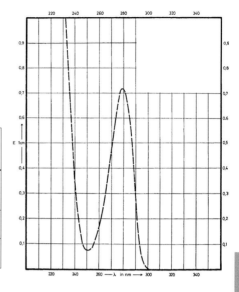

MW: 197.2

solvent symbol	methanol ——	water –·–·–	0.1M HCl – – – –	0.1M NaOH ······
absorption maximum			279 nm	
$E_{1cm}^{1\%}$			141	
ε			2780	

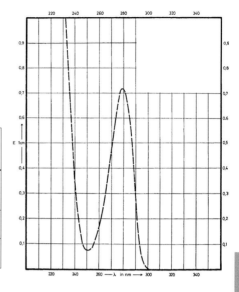

L – N

C_{max}-value	protein binding (PB)	elimination half life $t_{1/2}$	pKa-value	Ionisation MS	Parent Ion / Fragment Ion(s)
250 mg oral → 1400 ng/ml (usual 200 – 2500 ng/ml)	<5 %	1 – 3 h	2.3/8.7/ 9.7/13.4	pos	198 / 152/139/181
				neg	196 / 135/109

detection limits	LOD	precipitation	LLE	SPE
UV	1 ng (279 nm)			
FL				26 ng/ml [1]
ECD	0.1 ng	20 ng/ml [2]		
MS/MS				

(1) HPLC-Fluor (280/325 nm) SPE 26 ng/ml
(2) J Chromatogr B 802 (2004) 299-305, precipitation (HClO$_4$); HPLC-ECD 20 ng/ml

Lornoxicam

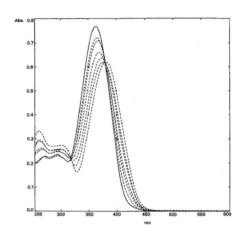

MW: 371.8

no UV
no E1/1

C_{max}-value	protein binding (PB)	elimination half life $t_{1/2}$	pKa-value	Ionisation MS	Parent Ion
					Fragment Ion(s)
4 mg oral → 2 µg/ml	99 %	3 – 4 h	3.0	pos	372
					121
				neg	370
					306/186/171/150

detection limits	LOD	precipitation	LLE	SPE
UV	1 ng (371 nm)			3 ng/ml [1] / 100 ng/ml [2]
FL				
ECD			5 ng/ml [3]	
MS/MS				

(1) HPLC-UV (371 nm), SPE 3 ng/ml
(2) J Chromatogr B 707 (1998) 151-159, SPE, HPLC-UV (372 nm) 100 ng/ml
(3) J Chromatogr B 617 (1993) 105-110, extraction; HPLC-ECD 5 ng/ml

Mefenamic Acid

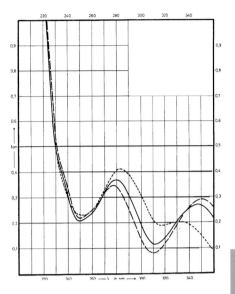

MW: 241.3

solvent symbol	methanol ———	water – · – · –	0.1M HCl – – – –	0.1M NaOH · · · · · ·
absorption maximum	346 nm 280 nm		350 nm 278 nm	332 nm 285 nm
$E_{1cm}^{1\%}$	268 367		288 344	202 409
ε	6470 8860		6950 8300	4860 9870

C_{max}-value	protein binding (PB)	elimination half life $t_{1/2}$	pKa-value	Ionisation MS	Parent Ion — Fragment Ion(s)
1000 mg → 10 µg/ml	99 %	3 – 4 h	4.2	pos	242 — 224/209/180
				neg	240 — 196/180

detection limits	LOD	precipitation	LLE	SPE
UV	1 ng (346 nm)		25 ng/ml [3]	
FL				
ECD				
MS/MS				10 ng/ml [2]

[2] Anal Bioanal Chem 384 (2006) 1501-1505, SPE; HPLC-MS 10ng/ml
[3] J Chromatogr B 800 (2004) 189-192, extraction; HPLC-UV (280 nm) 25 ng/ml

Metformin

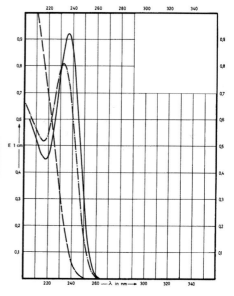

MW: 129.2

solvent symbol	methanol ——	water —·—·—	0.1M HCl ————	0.1M NaOH ······
absorption maximum	236 nm	232 nm		232 nm
$E_{1cm}^{1\%}$	908	800		787
ε	15030	13250		13030

C_{max}-value	protein binding (PB)	elimination half life $t_{1/2}$	pKa-value	Ionisation MS	Parent Ion / Fragment Ion(s)
500 mg oral → 1.6 µg/ml (usual 0.1 – 1.3 µg/ml)	<5 %	3 – 6 h	2.8/11.5	pos	130 / 60/71/85

detection limits	LOD	precipitation	LLE	SPE
UV	1 ng (236 nm)		16 ng/ml [3]	
FL				
ECD				
MS/MS		10 ng/ml [1] / 20 ng/ml [2]		

(1) HPLC-MS/MS ACN-precipitation 10 ng/ml
(2) J Chromatogr B 852 (2007) 308-316, ACN-precipitation; HPLC-MS/MS 20 ng/ml
(3) J. Chromatogr. B 824 (2005) 319-322, extraction; HPLC-UV 16 ng/ml

5-Methoxypsoralen + 8-Methoxypsoralen

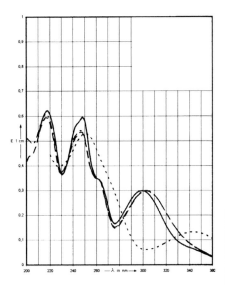

MW: 216.2

solvent symbol	methanol ———	water –·–·–	0.1M HCl – – – –	0.1M NaOH ······
absorption maximum	299 nm 248 nm 218 nm	302 nm 247 nm 217 nm	303 nm 247 nm 217 nm	345 nm 250 nm
$E_{1cm}^{1\%}$	567 1121 1158	558 1003 1105	559 1000 1105	245 984
ε	12200 24400 25000	12100 21700 23900	12100 21600 23900	5300 21300

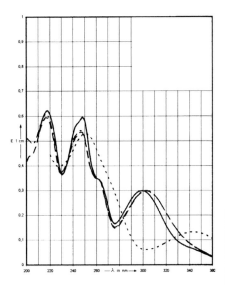

C_{max}-value	protein binding (PB)	elimination half life $t_{1/2}$	pKa-value	Ionisation MS	Parent Ion ——— Fragment Ion(s)
40 mg oral → 50 ng/ml 30 mg oral → 300 ng/ml	>90 %	2.3 h 3 h		pos	217 ——— 174

detection limits	LOD	precipitation	LLE	SPE
UV	1 ng (310 nm)		2 ng/ml [1]	1.5 ng/ml [2]
FL				
ECD				
MS/MS				

[1] Arzneim Forsch 32 (1982) 1338-1341, extraction; HPLC-UV (310 nm) 2 ng/ml
[2] Clin Chem 36 (1990) 1956-1957, SPE; HPLC-UV (300 nm) 1.5 ng/ml

Metoprolol

MW: 267.4

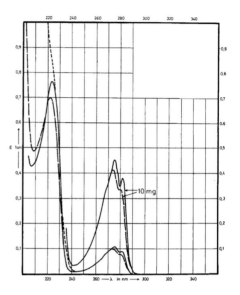

solvent symbol	methanol ———	water —·—·—	0.1M HCl —————	0.1M NaOH ······
absorption maximum	282 nm 276 nm 224 nm		274 nm 222nm	275 nm
$E^{1\%}_{1cm}$	38.0 45.2 304		40.9 278	41.3
ε	1590 1890 12690		1710 11610	1730

C_{max}-value	protein binding (PB)	elimination half life $t_{1/2}$	pKa-value	Ionisation MS	Parent Ion
100 mg oral → 150 ng/ml (usual 100 – 600 ng/ml)	12 %	2 – 7 h	9.7	pos	268
					77/116/103

detection limits	LOD	precipitation	LLE	SPE
UV	10 ng (276 nm)		20 ng/ml [3]	
FL			2 ng/ml [1]	
ECD				
MS/MS				10 ng/ml [2]

(1) Int J Clin Pharmacol Ther 34 (1996) 420-423, enant.; extraction, HPLC-Fluor (272/306 nm) 2 ng/ml
(2) Biomed Chromatogr 22 (2008) 702-711; HPLC-MS/MS 10 ng/ml
(3) Acta Chromatogr 19 (2007) 130-140; HPLC-UV (275 nm) 20 ng/ml

Metronidazole

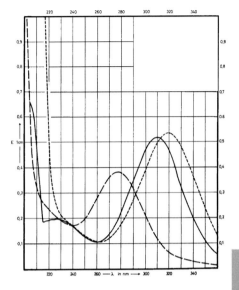

MW: 171.2

solvent symbol	methanol ———	water —·—·—	0.1M HCl ——————	0.1M NaOH ······
absorption maximum	310 nm		277 nm	319 nm
$E_{1cm}^{1\%}$	506		374	520
ε	8660		6400	8900

C_{max}-value	protein binding (PB)	elimination half life $t_{1/2}$	pKa-value	Ionisation MS	Parent Ion / Fragment Ion(s)
250 mg oral → 9 µg/ml (usual 10 – 30 µg/ml)	11%	8 h	2.5	pos	172 / 128/82

detection limits	LOD	precipitation	LLE	SPE
UV	1 ng (310 nm)	0.06 µg/ml [3]	0.4 µg/ml [1]	
FL				
ECD				
MS/MS				

(1) HPLC-UV (315 nm) 0.4 µg/ml extraction
(2) J Pharm Biomed Anal 38 (2005) 298-306, extraction; HPLC-MS/MS
(3) J Pharm Biomed Anal 37 (2005) 777-783, HClO$_4$-precipitation; HPLC-UV (315 nm) 60 ng/ml

Midazolam

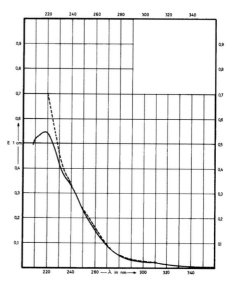

MW: 325.8

solvent symbol	methanol ———	water —·—·—	0.1M HCl —————	0.1M NaOH ······
absorption maximum				
$E_{1cm}^{1\%}$				
ε				

C_{max}-value	protein binding (PB)	elimination half life $t_{1/2}$	pKa-value	Ionisation MS	Parent Ion
					Fragment Ion(s)
20 mg oral → 260 ng/ml (usual 80 – 250 ng/ml)	95 %	2 h	4.3	pos	326
					291/244/249

detection limits	LOD	precipitation	LLE	SPE
UV	1 ng (220 nm)		8 ng/ml [3]	
FL				
ECD				
MS/MS	0.25 pg [1]		0.05 ng/ml [1] / 0.05 ng/ml [2]	

[1] HPLC-MS/MS extraction 0.05 ng/ml
[2] Rapid Commun Mass Spectrom 21 (2007) 1531-1540, extraction; HPLC-MS/MS 0.05 ng/ml
[3] J Chromatogr B 852 (2007) 571-577, extraction; HPLC-UV 8 ng/ml

Minocycline

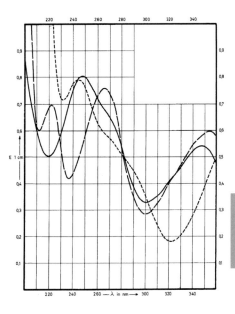

MW: 457.5

solvent symbol	methanol ———	water –·–·–	0.1M HCl –––––	0.1M NaOH ······
absorption maximum	348 nm 248 nm		354 nm 265 nm	385 nm 244 nm
$E_{1cm}^{1\%}$	256 385		290 370	335 383
ε	12630 19000		14300 18280	16560 18900

C_{max}-value	protein binding (PB)	elimination half life $t_{1/2}$	pKa-value	Ionisation MS	Parent Ion / Fragment Ion(s)
50 mg oral → 1.5 µg/ml	75 %	11 – 26 h	2.8/5.0/ 7.8/9.5	pos	458 441/352/283

detection limits	LOD	precipitation	LLE	SPE
UV	1 ng (348 nm)		30 ng/ml [1]	150 ng/ml [2]
FL				
ECD				
MS/MS				

[1] J Chromatogr A 812 (1998) 339-342, extraction; HPLC-UV (350 nm) 30 ng/ml
[2] Chromatographia 65 (2007) 277-281, SPE; HPLC-UV (350 nm) 150 ng/ml

Montelukast

MW: 586.2

no UV
no E1/1

C_{max}-value	protein binding (PB)	elimination half life $t_{1/2}$	pKa-value	Ionisation MS	Parent Ion
					Fragment Ion(s)
10 mg oral → 400 ng/ml	>99 %	2.7 – 5.5 h	9.5	pos	586
					568

detection limits	LOD	precipitation	LLE	SPE
UV	1 ng			
FL	0.1 ng [1]	5 ng/ml [1] / 5 ng/ml [2]		
ECD				
MS/MS		0.25 ng/ml [3]		

[1] HPLC-Fluor (350/410 nm) 5 ng/ml ACN-precipitation
[2] J Chromatogr B 869 (2008) 38-44, ACN-precipitation; HPLC-Fluor (350/450 nm) 5 ng/ml
[3] J Chromatogr B 858 (2007) 282-286, ACN-precipitation; HPLC-MS/MS 0.25 ng/ml

Nabilone

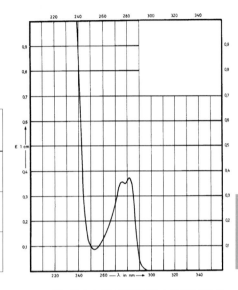

MW: 372.5

solvent symbol	methanol ———	water —·—·—	0.1M HCl —————	0.1M NaOH ⋯⋯
absorption maximum	282 nm 276 nm			
$E_{1cm}^{1\%}$	37 35			
ε	1360 1300			

C_{max}-value	protein binding (PB)	elimination half life $t_{1/2}$	pKa-value	Ionisation MS	Parent Ion ——— Fragment Ion(s)
2 mg oral → 2 ng/ml		2 h	7.1	pos	373 ——— 247

detection limits	LOD	precipitation	LLE	SPE
UV	10 ng (282 nm)			
FL				
ECD				
MS/MS	2 pg [1]		50 pg/ml [1]	

(1) Extraction; HPLC-MS/MS 50 pg/ml

Neomycin

neamine

neobiosamine B

Neomycin B   MW: 614.6

no UV
no E1/1

C_{max}-value	protein binding (PB)	elimination half life $t_{1/2}$	pKa-value	Ionisation MS	Parent Ion
					Fragment Ion(s)
2 mg oral → 2 ng/ml		4 h		pos (2++)	308
					454

detection limits	LOD	precipitation	LLE	SPE
UV	<			
FL				
ECD				
MS/MS	5 pg	0.2 µg/ml [1]		

(1) J Pharm Biomed Anal 43 (2007) 691-700, ACN-precipitation; HPLC-MS/MS 0.2 µg/ml

Nicorandil

MW: 211.2

no UV
no E1/1

C_{max}-value	protein binding (PB)	elimination half life $t_{1/2}$	pKa-value	Ionisation MS	Parent Ion
					Fragment Ion(s)
20 mg oral → 250 ng/ml	few	1 h	0.4	pos	212
					136

detection limits	LOD	precipitation	LLE	SPE
UV	10 ng (230 nm)		10 ng/ml [2]	
FL				
ECD				
MS/MS			1 ng/ml [1]	

[1] HPLC-MS/MS 1 ng/ml extraction
[2] J Pharm Biomed Anal 21 (1999) 175-178, extraction; HPLC-UV (230 nm) 10 ng/ml

Nifedipine

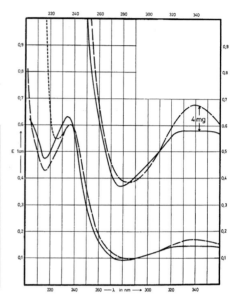

MW: 346.3

solvent symbol	methanol ———	water —·—·—	0.1M HCl —————	0.1M NaOH ······
absorption maximum	340 nm 235 nm		338 nm 238 nm	340 nm 238 nm
$E_{1cm}^{1\%}$	145 624		165 595	165 592
ε	5010 21590		5740 20600	5740 20510

C_{max}-value	protein binding (PB)	elimination half life $t_{1/2}$	pKa-value	Ionisation MS	Parent Ion
10 mg oral → 80 ng/ml (usual 10 – 200 ng/ml)	96 %	2 – 6 h	2.5	pos	347
					Fragment Ion(s)
					254/195/167

detection limits	LOD	precipitation	LLE	SPE
UV	1 ng (235 nm)		1 ng/ml [1]	5 ng/ml [2]
FL	<			
ECD				
MS/MS				

(1) Chromatographia 25 (1998) 919-922, extraction; HPLC-UV (235 nm) 1 ng/ml
(2) J Pharm Biom Anal 32 (2003) 1213-1218, SPE; HPLC-UV 5 ng/ml

Nitrofurantoin

MW: 238.2

solvent symbol	methanol ———	water —·—·—	0.1M HCl – – – –	0.1M NaOH ······
absorption maximum	357 nm 269 nm		266 nm	281 nm
$E_{1cm}^{1\%}$	669 427		567	568
ε	15940 10170		13510	13530

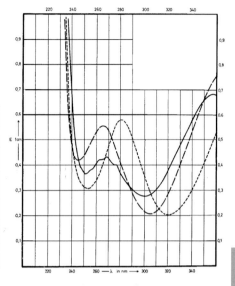

C_{max}-value	protein binding (PB)	elimination half life $t_{1/2}$	pKa-value	Ionisation MS	Parent Ion ——— Fragment Ion(s)
100 mg oral → 1 µg/ml	60 %	0.5 – 1 h	7.2	neg	237 ———

detection limits	LOD	precipitation	LLE	SPE
UV	1 ng (357 nm)		10 ng/ml [2]	
FL				
ECD				
MS/MS				

(2) J Chromatogr A 729 (1996) 251-258, extraction; HPLC-UV (370 nm) 10 ng/ml

Norfloxacin

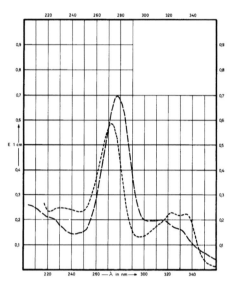

MW: 319.3

solvent symbol	methanol ———	water —·—·—	0.1M HCl —————	0.1M NaOH ······
absorption maximum			315 nm 277 nm	335 nm 323 nm 272 nm
$E_{1cm}^{1\%}$			383 1350	430 440 1130
ε			12200 43100	13700 14000 36100

C_{max}-value	protein binding (PB)	elimination half life $t_{1/2}$	pKa-value	Ionisation MS	Parent Ion / Fragment Ion(s)
400 mg oral → 1.3 µg/ml	14 %	3 – 4 h		pos	320 / 302/233/276

detection limits	LOD	precipitation	LLE	SPE
UV	0.1 ng (277 nm)			
FL	0.1 ng/ml [1]	25 ng/ml [1]		
ECD				
MS/MS				

[1] J Chromatogr A 812 (1998) 381-385, ACN-precipitation; HPLC-Fluor (300/450 nm) 25 ng/ml

Omeprazole

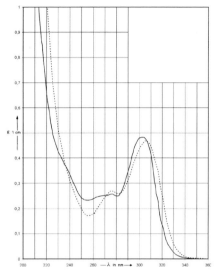

MW: 345.4

solvent symbol	methanol ———	water – · – · –	0.1M HCl – – – –	0.1M NaOH · · · · · ·
absorption maximum	302 nm	Decom-position observed	Decom-position observed	305 nm 276 nm
$E_{1cm}^{1\%}$	473			452 262
ε	16300			15600 9050

C_{max}-value	protein binding (PB)	elimination half life $t_{1/2}$	pKa-value	Ionisation MS	Parent Ion ——— Fragment Ion(s)
20 mg oral → 480 ng/ml (usual maximum 800 – 4400 ng/ml)	95 %	0.5 – 3 h	2.2	pos	345 ——— 297/149/282

detection limits	LOD	precipitation	LLE	SPE
UV	1 ng (302 nm)		2 ng/ml [3]	15 ng/ml [1]
FL				
ECD				
MS/MS				1 ng/ml [2]

(1) SPE; HPLC-UV 15 ng/ml
(2) J Liqu Chromatogr Relat Technol 30 (2007) 1797-1810, SPE; HPLC-MS/MS (APCI) 1 ng/ml
(3) J Chromatogr 844 (2006) 314-321, extraction; HPLC-UV 2 ng/ml

Oxazepam

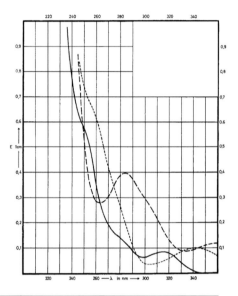

MW: 286.7

solvent symbol	methanol ———	water —·—·—	0.1M HCl ————	0.1M NaOH ······
absorption maximum	315 nm		360 nm 283 nm	344 nm
$E_{1cm}^{1\%}$	84		120 400	102
ε	2410		3440 11470	2920

C_{max}-value	protein binding (PB)	elimination half life $t_{1/2}$	pKa-value	Ionisation MS	Parent Ion ——— Fragment Ion(s)
40 mg oral → 600 ng/ml (usual 0.5 – 2.0 µg/ml)	95 %	4 – 25 h	1.7/11.6	pos	287 ——— 241/269/104

detection limits	LOD	precipitation	LLE	SPE
UV	1 ng (230 nm)		5 ng/ml [1] / 2 ng/ml [2]	
FL				
ECD				
MS/MS			5 ng/ml [3]	

[1] Extraction; HPLC-UV 5 ng/ml
[2] Talanta 75 (2008), 671-676, extraction; HPLC-UV (230 nm) 2 ng/ml
[3] J Pharm Biomed Anal 41 (2006) 1135-1145, extraction; HPLC-MS 5 ng/ml

Paclitaxel

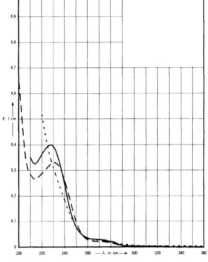

MW: 853.9

solvent symbol	methanol ———	water —·—·—	0.1M HCl —————	0.1M NaOH ······
absorption maximum	228 nm		231 nm	
$E_{1cm}^{1\%}$	403		333	
ε	34400		28400	

C_{max}-value	protein binding (PB)	elimination half life $t_{1/2}$	pKa-value	Ionisation MS	Parent Ion
					Fragment Ion(s)
50 – 500 ng/ml	95 – 98 %	3 – 50 h	3.3	pos	854
					286

detection limits	LOD	precipitation	LLE	SPE
UV	1 ng (302 nm)		250 ng/ml [2] / 10 ng/ml [3]	
FL				
ECD				
MS/MS			2 ng/ml [1] / 0.2 ng/ml [4]	

(1) Extraction; HPLC-MS/MS 2 ng/ml
(2) J Liqu. Chromatogr Relat Technol 31 (2008) 1478-1491, extraction; HPLC-UV (228 nm) 0.25 µg/ml
(3) Biomed Chrom 21 (2007) 310-317, extraction; HPLC-UV (230 nm) 10 ng/ml
(4) J Chrom Sci. 44 (2006) 266-271, extraction; HPLC-MS/MS 0.2 ng/ml

Pantoprazole

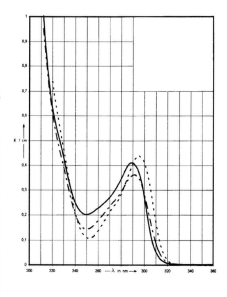

MW: 383.4

solvent symbol	methanol ———	water —·—·—	0.1M HCl – – – –	0.1M NaOH ......
absorption maximum	289 nm	291 nm	Decomposition observed	295 nm
$E_{1cm}^{1\%}$	391	346		418
ε	16600	14700		17700

C_{max}-value	protein binding (PB)	elimination half life $t_{1/2}$	pKa-value	Ionisation MS	Parent Ion ——— Fragment Ion(s)
40 mg oral → 1.2 µg/ml (usual maximum 1.1 – 3.1 µg/ml)	98 %	1 h	1.7	pos	384 ——— 200

detection limits	LOD	precipitation	LLE	SPE
UV	1 ng (289 nm)	250 ng/ml [2]		30 ng/ml [1a]
FL				
ECD				
MS/MS		10 ng/ml [1b]		

(1a) SPE; HPLC-UV 30 ng/ml
(1b) ACN-precipitation; HPLC-MS/MS 10 ng/ml
(2) Pharmazie 61 (2006) 586-589, ACN-precipitation; HPLC-UV (272 nm) 250 ng/ml

Paracetamol

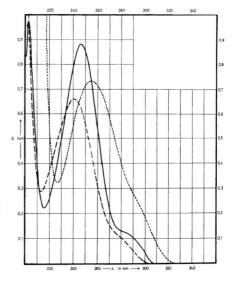

MW: 151.2

solvent symbol	methanol ——	water —·—·—	0.1M HCl —————	0.1M NaOH ······
absorption maximum	247 nm		240 nm	255 nm
$E_{1cm}^{1\%}$	850		642	710
ε	12850		9710	10740

C_{max}-value	protein binding (PB)	elimination half life $t_{1/2}$	pKa-value	Ionisation MS	Parent Ion ——— Fragment Ion(s)
10 – 20 µg/ml (usual therapeutic range toxic/poisonous >300 µg/ml) 500 mg oral → 5 µg/ml	<5 %	1.5 – 3 h	9.5	pos	152 ——— 110/65/93

detection limits	LOD	precipitation	LLE	SPE
UV	1 ng (247 nm)			1250 ng/ml [3]
FL				
ECD	0.1 ng [1]	50 ng/ml [1]		
MS/MS			50 ng/ml [2]	

(1) ACN-precipitation; HPLC-ECD 50 ng/ml
(2) Chromatographia, extraction; HPLC-MS 50 ng/ml
(3) J Pharm Biomed Anal 36 (2004) 737-741, SPE; HPLC-UV (220 nm) 1.25 µg/ml

O

R

Paroxetine

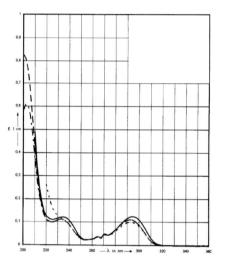

MW: 329.4

solvent symbol	methanol ———	water —·—·—	0.1M HCl – – – –	0.1M NaOH ······
absorption maximum	294 nm 235 nm	293 nm 233 nm	293 nm 233 nm	293 nm
$E_{1cm}^{1\%}$	125 121	110 112	110 112	101
ε	4700 4500	4100 4200	4100 4200	3800

C_{max}-value	protein binding (PB)	elimination half life $t_{1/2}$	pKa-value	Ionisation MS	Parent Ion — Fragment Ion(s)
25 mg oral → 15 ng/ml (usual 10 – 100 ng/ml)	95 %	21 h	4.7	pos	330 — 192/70/135

detection limits	LOD	precipitation	LLE	SPE
UV	1 ng (294 nm)		10 ng/ml [2]	
FL			7 ng/ml [3]	
ECD				
MS/MS	10 pg [1]		0.2 ng/ml [1]	

(1) HPLC-MS/MS 0.2 ng/ml extraction
(2) J Pharm Biomed Anal 44 (2007) 955-962, extraction; HPLC-UV (230 nm) 10 ng/ml
(3) J Liqu Chromatogr Relat Technol 30 (2007), 1641-1655, extraction; HPLC-Fluor (295/350 nm) 7 ng/ml

Penicillin V (Phenoxymethylpenicillin)

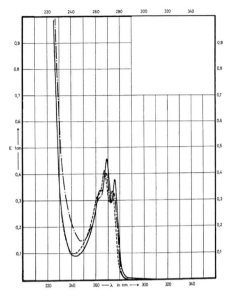

MW: 350.4

solvent symbol	methanol ———	water —·—·—	0.1M HCl –––––	0.1M NaOH ······
absorption maximum	275 nm 268 nm	275 nm 268 nm		273 nm 266 nm
$E_{1cm}^{1\%}$	29 35	25 31		26 32
ε	1130 1360	970 1200		1010 1240

C_{max}-value	protein binding (PB)	elimination half life $t_{1/2}$	pKa-value	Ionisation MS	Parent Ion / Fragment Ion(s)
700 mg oral → 7 µg/ml	80 %	0.5 h	2.7	pos	351 / 160/114
				neg	349 / 93/114

detection limits	LOD	precipitation	LLE	SPE
UV	10 ng (240 nm)			0.11 µg/ml [1] / 0.1 µg/ml [2] / 0.02 µg/ml [3]
FL				
ECD				
MS/MS				

(1) SPE; HPLC-UV (220 nm) 0.11 µg/ml
(2) J Chromatogr B 679 (1996) 129-135, SPE; HPLC-UV (269 nm) 0.1 µg/ml
(3) Chromatographia 53 (2001) 367-371, SPE + derivatization; HPLC-UV 0.02 µg/ml

O – R

Pentoxiphylline

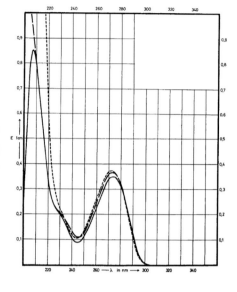

MW: 278.3

solvent symbol	methanol ——	water –·–·–	0.1M HCl – – – –	0.1M NaOH ······
absorption maximum	273 nm		274 nm	272 nm
$E_{1cm}^{1\%}$	342		359	365
ε	9520		9990	10160

C_{max}-value	protein binding (PB)	elimination half life $t_{1/2}$	pKa-value	Ionisation MS	Parent Ion ———— Fragment Ion(s)
600 mg oral retard → 100 ng/ml (usual 20 – 200 ng/ml)	<5 %	0.4 – 0.8 h	0.6	pos	279 ———— 181/99/138

detection limits	LOD	precipitation	LLE	SPE
UV	1 ng (273 nm)		10 ng/ml [1]	
FL				
ECD				
MS/MS				

[1] Extraction, HPLC-UV (273 nm) 10 ng/ml

Phenobarbital

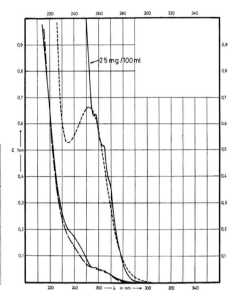

MW: 232.2

25 mg /100 ml

solvent symbol	methanol ————	water —·—·—	0.1M HCl —————	0.1M NaOH ......
absorption maximum	258 nm			252 nm
$E_{1cm}^{1\%}$	26			325
ε	600			7550

C_{max}-value	protein binding (PB)	elimination half life $t_{1/2}$	pKa-value	Ionisation MS	Parent Ion —————— Fragment Ion(s)
150 mg oral → 10 – 30 µg/ml (usual 10 – 40 µg/ml)	50 %	50 – 150 h	7.4	neg	231 —————— 144/188

detection limits	LOD	precipitation	LLE	SPE
UV	10 ng (258 nm)	1 µg/ml [3]		
FL				
ECD				
MS/MS				0.065 µg/ml [2]

(2) J Chromatogr B 857 (2007) 40-46, SPE; HPLC-MS 0.065 µg/ml
(3) Biomed Chromatogr 16 (2002) 19-24; ACN-precipitation; HPLC-UV (240 nm) 1 µg/ml

O – R

Phenylbutazone

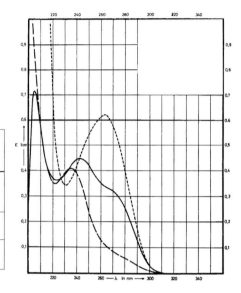

MW: 308.4

solvent symbol	methanol ——	water —·—·—	0.1M HCl ————	0.1M NaOH ······
absorption maximum	243 nm		235 nm	263 nm
$E_{1cm}^{1\%}$	482		440	669
ε	14 860		13 570	20 630

C_{max}-value	protein binding (PB)	elimination half life $t_{1/2}$	pKa-value	Ionisation MS	Parent Ion ———— Fragment Ion(s)
300 mg oral → 38 µg/ml (usual 50 – 150 µg/ml)	99 %	28 – 120 h	4.4	pos	309 ———— 160/77/92

detection limits	LOD	precipitation	LLE	SPE
UV	1 ng (243 nm)	250 ng/ml [2]	50 ng/ml [1]	
FL				
ECD				
MS/MS				

(1) Extraction; HPLC-UV 50 ng/ml
(2) J Chromatogr B 769 (2002) 119-126, ACN-precipitation; HPLC-UV 250 ng/ml

Phenytoin

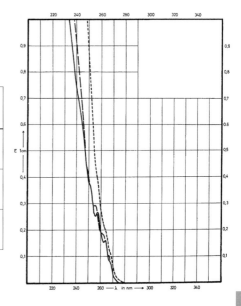

MW: 252.3

solvent symbol	methanol ———	water –·–·–	0.1M HCl – – – –	0.1M NaOH ......
absorption maximum	264 nm 258 nm			
$E_{1cm}^{1\%}$	16.0 26.8			
ε	400 680			

C_{max}-value	protein binding (PB)	elimination half life $t_{1/2}$	pKa-value	Ionisation MS	Parent Ion ——— Fragment Ion(s)
100 mg oral → 2.2 µg/ml → (usual 10 – 20 µg/ml)	90 %	7 – 60 h	8.3	pos	253 ——— 182/104
				neg	251 ——— 102/208

detection limits	LOD	precipitation	LLE	SPE
UV	10 ng (250 nm)	0.15 µg/ml [1] / 1 µg/ml [3]		0.065 µg/ml [2]
FL				
ECD				
MS/MS				

[1] HPLC-UV (220 nm) 0.15 µg/ml ACN-precipitation
[2] J Chromatogr B 857 (2007) 40-46, SPE; HPLC-UV 0.065 µg/ml
[3] Chromatographia 65 (2007) 267-275, MeOH precipitation; HPLC-UV (215 nm) 1 µg/ml

Pimelic Acid

MW: 160.2

no UV
no E1/1

C_{max}-value	protein binding (PB)	elimination half life $t_{1/2}$	pKa-value	Ionisation MS	Parent Ion
					Fragment Ion(s)
usual 2 – 10 ng/ml				neg	159
					97

detection limits	LOD	precipitation	LLE	SPE
UV	<			
FL	<			
ECD				
MS/MS			0.5 ng/ml [1]	

(1) Extraction; HPLC-MS/MS 0.5 ng/ml

Piracetam

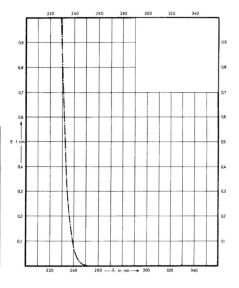

MW: 142.2

solvent symbol	methanol ———	water —·—·—	0.1M HCl —————	0.1M NaOH ······
absorption maximum				
$E_{1cm}^{1\%}$				
ε				

C_{max}-value	protein binding (PB)	elimination half life $t_{1/2}$	pKa-value	Ionisation MS	Parent Ion / Fragment Ion(s)
800 mg oral → 20 µg/ml	15 %	5.2 h		pos	143 / 126/98/70

detection limits	LOD	precipitation	LLE	SPE
UV	<	1 µg/ml [1] / 2 µg/ml [2]		
FL				
ECD				
MS/MS				

[1] J Pharm Biomed Anal 7 (1989) 913-916 precipitation; HPLC-UV 1 µg/ml
[2] J Biochem Biophys Meth. 69 (2007), 273-281, HClO$_4$-precipitation; HPLC-UV (200 nm) 2 µg/ml

O
–
R

Pirenzepine

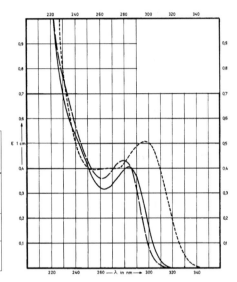

MW: 351.4

solvent symbol	methanol ———	water —.—.—	0.1M HCl — — —	0.1M NaOH ......
absorption maximum	284 nm		280 nm	297 nm
$E_{1cm}^{1\%}$	186		197	233
ε	7900		8360	9890

C_{max}-value	protein binding (PB)	elimination half life $t_{1/2}$	pKa-value	Ionisation MS	Parent Ion ——— Fragment Ion(s)
50 mg oral → 40 ng/ml	10 %	11 h	2.1/8.1	pos	352 ——— 113/70/252

detection limits	LOD	precipitation	LLE	SPE
UV	1 ng (284 nm)		5 ng/ml [2]	1 ng/ml [1]
FL				
ECD				
MS/MS				

(1) Arzneim Forsch 36 (1986) 1409-1412; HPLC-UV (283 nm) 1 ng/ml SPE
(2) J Chromatogr 375 (1986) 369-375, extraction; HPLC-UV (280 nm) 5 ng/ml

Piroxicam

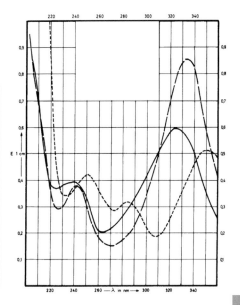

MW: 331.4

solvent symbol	methanol ———	water —·—·—	0.1M HCl – – – –	0.1M NaOH ······
absorption maximum	325 nm 240 nm		334 nm 241 nm	353 nm 286 nm 251 nm
$E_{1cm}^{1\%}$	556 368		813 361	475 300 395
ε	18430 12200		26940 11960	15740 9940 13090

C_{max}-value	protein binding (PB)	elimination half life $t_{1/2}$	pKa-value	Ionisation MS	Parent Ion ——— Fragment Ion(s)
20 mg oral → 2 µg/ml (usual 2 – 20 µg/ml)	99 %	30 – 60 h	2.6	pos	332 ——— 95/164/121
				neg	330 ——— 146/131/119

detection limits	LOD	precipitation	LLE	SPE
UV	1 ng (325 nm)	0.1 µg/ml [1] / 0.025 µg/ml [2]		
FL				
ECD				
MS/MS				

(1) ACN-precipitation; HPLC-UV (330 nm) 0.1 µg/ml
(2) Chromatographia 59 (2004) 555-560, ACN-precipitation; HPLC-UV (330 nm) 0.025 µg/ml

8-Prenylnaringenin

MW: 340.0

no UV
no E1/1

C$_{max}$-value		protein binding (PB)	elimination half life t$_{1/2}$	pKa-value	Ionisation MS	Parent Ion
						Fragment Ion(s)
					pos	341
						165/285

detection limits	LOD	precipitation	LLE	SPE
UV	1 ng			
FL				
ECD				
MS/MS	10 pg [1]		0.1 ng/ml [1]	

(1) Extraction; HPLC-MS/MS 0.1 ng/ml

Propafenone

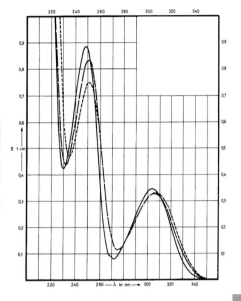

MW: 341.4

solvent symbol	methanol ———	water —·—·—	0.1M HCl —————	0.1M NaOH ······
absorption maximum	304 nm 248 nm		305 nm 251 nm	306 nm 251 nm
$E_{1cm}^{1\%}$	84 217		81 205	81 184
ε	3170 8180		3060 7740	3060 6960

C_{max}-value	protein binding (PB)	elimination half life $t_{1/2}$	pKa-value	Ionisation MS	Parent Ion ——— Fragment Ion(s)
(usual 250 – 1000 ng/ml)	85 – 95 %	5 bzw. 20 h	3.4	pos	342 ——— 116/72/324

detection limits	LOD	precipitation	LLE	SPE
UV	1 ng (248 nm)	10 ng/ml [2]		
FL				
ECD				
MS/MS				3 ng/ml [3]

[2] Anal Sci 20 (2004) 1307-1311, ACN-precipitation; HPLC-UV (210 nm) 10 ng/ml
[3] J Chromatogr B 748 (2000) 113-123, SPE; HPLC-MS 3 ng/ml

Propranolol

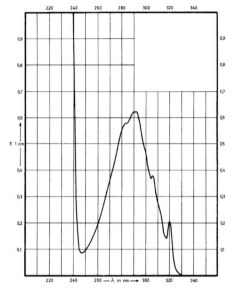

MW: 259.3

solvent symbol	methanol ———	water –·–·–	0.1M HCl ––––	0.1M NaOH ······
absorption maximum	320 nm 292 nm			
$E^{1\%}_{1cm}$	80 236			
ε	2070 6110			

C_{max}-value	protein binding (PB)	elimination half life $t_{1/2}$	pKa-value	Ionisation MS	Parent Ion
					Fragment Ion(s)
40 mg oral → 30 ng/ml → >20 ng/ml active (usual 50 – 300 ng/ml)	90 %	2 – 6 h	9.5	pos	260 116/183/155

detection limits	LOD	precipitation	LLE	SPE
UV	1 ng (292 nm)			
FL				
ECD				
MS/MS				10 ng/ml [2]

(2) Biomed Chromatogr 22 (2008) 702-711, SPE; HPLC-MS/MS 10 ng/ml

Propyphenazone

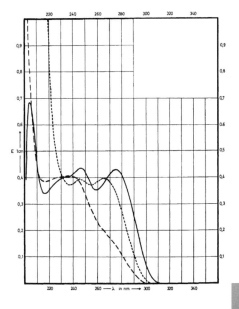

MW: 230.3

solvent symbol	methanol ———	water —·—·—	0.1M HCl —————	0.1M NaOH ······
absorption maximum	275 nm 246 nm		240 nm	265 nm 245 nm
$E_{1cm}^{1\%}$	420 425		400	390 385
ε	9670 9790		9210	8980 8870

C_{max}-value	protein binding (PB)	elimination half life $t_{1/2}$	pKa-value	Ionisation MS	Parent Ion ——— Fragment Ion(s)
300 – 1500 mg daily (usual 1.5 – 3.5 µg/ml)	10 %	1 – 1.5 h	2.0	pos	231 ——— 189/56/ 201/77

detection limits	LOD	precipitation	LLE	SPE
UV	1 ng (275 nm)	100 ng/ml [2]	50 ng/ml [1]	
FL				
ECD				
MS/MS				

[1] Extraction; HPLC-UV 50 ng/ml
[2] J Chromatogr 577 (1992) 387-390, ACN-precipitation; HPLC-UV (270 nm) 100 ng/ml

Roflumilast

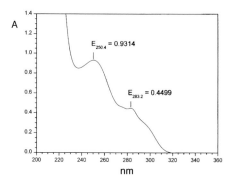

W: 403.2

no UV
no E1/1

C_{max}-value	protein binding (PB)	elimination half life $t_{1/2}$	pKa-value	Ionisation MS	Parent Ion
					Fragment Ion(s)
0.5 mg oral → 7 ng/ml	99 %	10 h	3.5	pos	403
					187

detection limits	LOD	precipitation	LLE	SPE
UV	10 ng			
FL				
ECD				
MS/MS			40 pg/ml [1]	

(1) Extraction; HPLC-MS/MS 40 pg/ml

Salbutamol

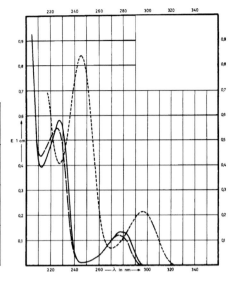

MW: 239.3

solvent symbol	methanol ———	water –·–·–	0.1M HCl – – – –	0.1M NaOH ······
absorption maximum	278 nm 227 nm	276 nm 225 nm	276 nm 225 nm	295 nm 245 nm
$E_{1cm}^{1\%}$	68 289	61 272	61 272	110 423
ε	3930 16700	3500 15700	3500 15700	6330 24400

C_{max}-value	protein binding (PB)	elimination half life $t_{1/2}$	pKa-value	Ionisation MS	Parent Ion ——— Fragment Ion(s)
200 µg inhal. → 500 pg/ml (usual 1 – 20 ng/ml)	low	2 – 7 h	9.3/10.3	pos	240 ——— 148/166/222

detection limits	LOD	precipitation	LLE	SPE
UV	1 ng (227 nm)			
FL			0.8 ng/ml [3]	10 ng/ml [2]
ECD				
MS/MS				0.02 ng/ml [1]

(1) HPLC-MS/MS 0.020 ng/ml SPE
(2) J Chromatogr A 987 (2003), 257-267, SPE; HPLC-Fluor 10 ng/ml
(3) Biomed Chromatogr 14 (2000), 1-5, extraction; HPLC-Fluor (230/306 nm) 0.8 ng/ml

S – V

Salicylic Acid

MW: 138.1

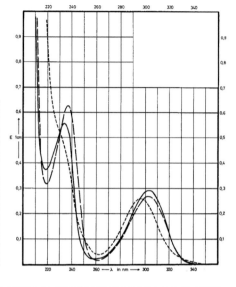

solvent symbol	methanol ———	water —·—·—	0.1M HCl —————	0.1M NaOH ······
absorption maximum	302 nm 234 nm		303 nm 237 nm	296 nm
$E_{1cm}^{1\%}$	285 547		262 613	254
ε	3940 7550		3620 8470	3510

C_{max}-value	protein binding (PB)	elimination half life $t_{1/2}$	pKa-value	Ionisation MS	Parent Ion / Fragment Ion(s)
therapeutic 50 – 300 µg/ml	50 – 90 %	2 – 4 h	3.0/13.4	neg	137 / 93/65/75

detection limits	LOD	precipitation	LLE	SPE
UV	1 ng (302 nm)			38 ng/ml [3]
FL	10 pg [1]	1 ng/ml [1]		
ECD				
MS/MS		20 ng/ml [2]		

(1) HPLC-Fluor (319/399 nm) 1 ng/ml ACN-precipitation
(2) Chromatographia 67 (2008) 583-590, ACN-precipitation; HPLC-MS/MS 20 ng/ml
(3) Biomed Chromatogr 21 (2007) 221-224, SPE; HPLC-UV 38 ng/ml

Salmeterol

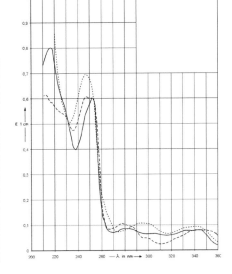

MW: 415.6

solvent symbol	methanol ———	water —·—·—	0.1M HCl —————	0.1M NaOH ······
absorption maximum	343 nm 283 nm 252 nm	343 nm 250 nm 217 nm	341 nm 278 nm 246 nm	341 nm 295 nm 246 nm
$E_{1cm}^{1\%}$	77.4 83.9 593	74.6 567 724	78.1 104 594	91.8 106 687
ε	4670 5060 35800	4500 34200 43600	4710 6240 35800	5540 6380 41400

C_{max}-value	protein binding (PB)	elimination half life $t_{1/2}$	pKa-value	Ionisation MS	Parent Ion ——— Fragment Ion(s)
50 – 200 pg/ml	93 %		4.2	pos	416 ——— 398/232/91

detection limits	LOD	precipitation	LLE	SPE
UV	1 ng (252 nm)			
FL				
ECD	0.1 ng			
MS/MS				2 pg/ml [1] / 2.5 pg/ml [2]

(1) SPE; HPLC-MS/MS 2 pg/ml
(2) J Chromatogr B 876 (2008) 163-169, SPE; HPLC-MS/MS 2.5 pg/ml

Silibinin = Silymarin

MW: 482.4

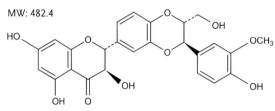

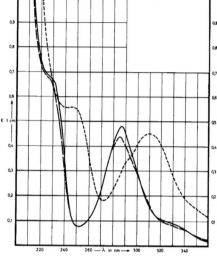

solvent symbol	methanol ———	water —·—·—	0.1M HCl —————	0.1M NaOH ······
absorption maximum	287 nm		286 nm	310 nm
$E_{1cm}^{1\%}$	450		411	423
ε	21700		19800	20400

C_{max}-value	protein binding (PB)	elimination half life $t_{1/2}$	pKa-value	Ionisation MS	Parent Ion ——— Fragment Ion(s)
100 mg oral → 110 ng/ml free / total 2500 ng/ml (Glucuronid)		2.5 h	1.9	neg	481 ——— 301/125

detection limits	LOD	precipitation	LLE	SPE
UV	1 ng (287 nm)		2.5 ng/ml [1]	
FL				
ECD				
MS/MS				2 ng/ml [2]

[1] J Liqu Chromatogr 16 (1993) 2777-2789, extraction; HPLC-UV (285 nm) 2.5 ng/ml
[2] J Chromatogr B 862 (2008) 51-57; HPLC-MS/MS 2 ng/ml

Temazepam

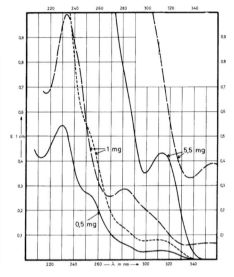

MW: 300.7

solvent symbol	methanol ———	water —·—·—	0.1M HCl ————	0.1M NaOH ······
absorption maximum	314 nm 230 nm		358 nm 284 nm 234 nm	310 nm
$E_{1cm}^{1\%}$	76 1080		68 283 980	81
ε	2300 32480		2050 8500 29420	2440

C_{max}-value	protein binding (PB)	elimination half life $t_{1/2}$	pKa-value	Ionisation MS	Parent Ion ——— Fragment Ion(s)
20 mg oral → 700 ng/ml (usual 200 – 800 ng/ml)	97 %	3 – 38 h	1.6	pos	301 ——— 225/177/193

detection limits	LOD	precipitation	LLE	SPE
UV	0.1 ng (230 nm)		5 ng/ml [1] / 2 ng/ml [2]	
FL				
ECD				
MS/MS				0.3 ng/ml [3]

[1] Extraction, HPLC-UV 5 ng/ml
[2] Talanta 75 (2008) 671-676; extraction, HPLC-UV (230 nm) 2 ng/ml
[3] Pharm Biomed Anal 26 (2001) 321-330; SPE, HPLC-MS/MS 0.3 ng/ml

Theophylline

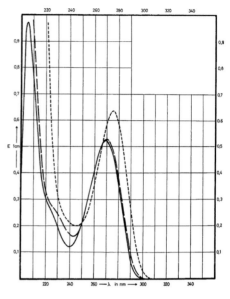

MW: 180.2

solvent symbol	methanol ———	water –·–·–	0.1M HCl ––––	0.1M NaOH ······
absorption maximum	269 nm		270 nm	275 nm
$E_{1cm}^{1\%}$	501		538	632
ε	9930		10660	12530

C_{max}-value	protein binding (PB)	elimination half life $t_{1/2}$	pKa-value	Ionisation MS	Parent Ion — Fragment Ion(s)
(usual 8 – 20 µg/ml)	56 %	3 – 13 h	<1/8.6	pos	181 — 124/81/69

detection limits	LOD	precipitation	LLE	SPE
UV	1 ng (269 nm)	0.1 µg/ml [1] / 0.1 µg/ml [2]		
FL				
ECD				
MS/MS	2 pg	0.1 µg/ml [3]		

[1] Precipitation, HPLC-UV 0.1 µg/ml
[2] J Chromatogr B 848 (2007) 271-276; HPLC-UV (272 nm) 0.1 µg/ml
[3] Biomed Chrom 19 (2005) 643-648; ACN-precipitation HPLC-MS/MS 0.1 µg/ml

Tolperisone (+Enantiomers)

H₃C ... (structure) ... CH₃ ... N

MW: 245.4

no UV
no E1/1

C$_{max}$-value	protein binding (PB)	elimination half life t$_{1/2}$	pKa-value	Ionisation MS	Parent Ion
					Fragment Ion(s)
300 mg oral → 300 ng/ml	90 %	2.5 h	3.7	pos	246
					55

detection limits	LOD	precipitation	LLE	SPE
UV	10 ng			
FL				
ECD				
MS/MS			2 ng/ml [1]	

(1) Extraction; HPLC-MS/MS 0.2 ng/ml (enantiomer separation)

Tramadol

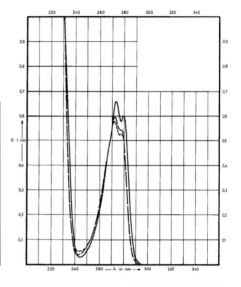

H₃C—N—CH₃

H₃CO

OH

MW: 263.4

solvent symbol	methanol ———	water —·—·—	0.1M HCl ————	0.1M NaOH ······
absorption maximum	279 nm 273 nm		272 nm	272 nm
$E_{1cm}^{1\%}$	59 66		59	58
ε	1780 1960		1780	1720

C_{max}-value	protein binding (PB)	elimination half life $t_{1/2}$	pKa-value	Ionisation MS	Parent Ion
50 mg retard → 70 ng/ml (usual 300 – 900 ng/ml)	<5 %	6 h	8.3	pos	264 —— 58/42

detection limits	LOD	precipitation	LLE	SPE
UV	10 ng (273 nm)			
FL			4 ng/ml [1] / 4 ng/ml [2]	
ECD				
MS/MS		2 ng/ml [3]		

(1) Extraction HPLC-Fluor (220/305 nm) 4 ng/ml
(2) J Biochem Biophys Meth 70 (2008) 1304-1312, extraction; HPLC-fluorescence 4 ng/ml
(3) Talanta 75 (2008) 1104-1109, HClO₄-precipitation; HPLC-MS 2 ng/ml

Triamterene

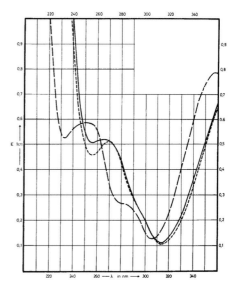

MW: 253.3

solvent symbol	methanol ———	water —·—·—	0.1M HCl —————	0.1M NaOH ······
absorption maximum	266 nm		357 nm 250 nm	270 nm
$E_{1cm}^{1\%}$	568		855 634	559
ε	14390		21660 16060	14160

C_{max}-value	protein binding (PB)	elimination half life $t_{1/2}$	pKa-value	Ionisation MS	Parent Ion ——— Fragment Ion(s)
50 mg oral → 60 ng/ml	45 – 70 %	2 – 4 h	6.2	pos	254 ——— 237/104/195

detection limits	LOD	precipitation	LLE	SPE
UV	0.1 ng (365 nm)			
FL	0.1 pg	1 ng/ml [1]		
ECD				
MS/MS				

(1) Dilution HPLC-Fluor (360/436 nm) 1 ng/ml

Tryptophan

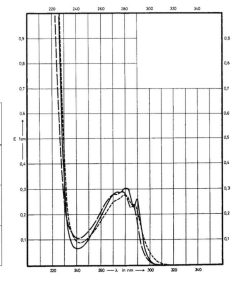

MW: 204.2

solvent symbol	methanol ——	water —·—·—	0.1M HCl — — — —	0.1M NaOH ······
absorption maximum	290 nm 280 nm		286 nm 278 nm	288 nm 280 nm
$E_{1cm}^{1\%}$	259 303		234 290	238 275
ε	5290 6190		4780 5920	4860 5620

C_{max}-value	protein binding (PB)	elimination half life $t_{1/2}$	pKa-value	Ionisation MS	Parent Ion
					Fragment Ion(s)
usual up to 16 µg/ml			2.4/9.4	pos	205 ——— 188/146/118

detection limits	LOD	precipitation	LLE	SPE
UV	1 ng (290 nm)			
FL	1 ng	13 µg/ml [2]		
ECD		1 µg/ml [1]		
MS/MS		0.1 µg/ml [3]		

(1) HPLC-ECD 1 µg/ml dilution
(2) Clin Chem 48 (2002), TCA-precipitation; HPLC-Fluor (286/366 nm) 13 µg/ml
(3) Anal Chem 74 (2002) 2034-2040; HPLC-MS/MS 0.1 µg/ml

Valnemulin

MW: 564.8

no UV
no E1/1

C$_{max}$-value	protein binding (PB)	elimination half life t$_{1/2}$	pKa-value	Ionisation MS	Parent Ion
					Fragment Ion(s)
			4.3	pos	566

detection limits	LOD	precipitation	LLE	SPE
UV	<			
FL			20 ng/ml [1]	
ECD				
MS/MS	10 pg			

(1) Extraction/derivatization, HPLC-Fluor (420/490 nm) 20 ng/ml

Verapamil

MW: 454.6

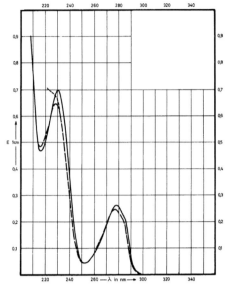

solvent symbol	methanol ———	water —·—·—	0.1M HCl ————	0.1M NaOH ······
absorption maximum	279 nm 230 nm		278 nm 228 nm	278 nm
$E_{1cm}^{1\%}$	123 335		118 313	113
ε	6050 16400		5800 15400	5550

C_{max}-value	protein binding (PB)	elimination half life $t_{1/2}$	pKa-value	Ionisation MS	Parent Ion — Fragment Ion(s)
40 mg oral → 60 ng/ml >100 ng/ml active (usual 50 – 750 ng/ml)	90 %	2 – 7 h	4.8	pos	455 — 165/150/303

detection limits	LOD	precipitation	LLE	SPE
UV	1 ng (230 nm)			10 ng/ml [2]
FL	0.1 ng [1]		3 ng/ml [1]	10 ng/ml [3]
ECD				
MS/MS				

(1) Extraction; HPLC-Fluor (275/320 nm) 3 ng/ml
(2) J Biochem Biophys Meth 70 (2008) 1297-1303, SPE; HPLC-UV (200 nm) 10 ng/ml
(3) Pharm Biomed Anal 37 (2005) 405-410, SPE; HPLC-Fluor 10 ng/ml

Vitamin B1 (Thiamine)

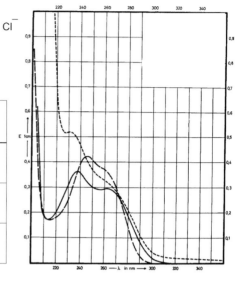

MW: 300.8

solvent symbol	methanol ———	water —.—.—	0.1M HCl – – – –	0.1M NaOH ......
absorption maximum	262 nm 238 nm		246 nm	232 nm
$E^{1\%}_{1cm}$	286 352		413	505
ε	9650 11870		13930	17030

C_{max}-value	protein binding (PB)	elimination half life $t_{1/2}$	pKa-value	Ionisation MS	Parent Ion —————— Fragment Ion(s)
200 mg oral → 30 ng/ml 2 – 13 ng/ml normal level			4.8	pos	265 —————— 122/144/81

detection limits	LOD	precipitation	LLE	SPE
UV	1 ng (262 nm)			
FL		2 ng/ml [1]		
ECD				
MS/MS		1 ng/ml [2]		

[1] J Pharm Sci 82 (1993) 56-59, derivatization; HPLC-Fluor (365/435 nm) 2 ng/ml
[2] Chromatographia 67 (2008) 583-590, precipitation; HPLC-MS/MS 1 ng/ml

Vitamin B2 (Riboflavin)

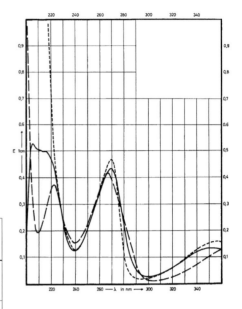

MW: 376.4

solvent symbol	methanol ———	water —·—·—	0.1M HCl —————	0.1M NaOH ······
absorption maximum	352 nm 269 nm		267 nm 223 nm	356 nm 270 nm
$E_{1cm}^{1\%}$	274 890		854 760	325 959
ε	10310 33500		32140 28610	12230 36100

C_{max}-value	protein binding (PB)	elimination half life $t_{1/2}$	pKa-value	Ionisation MS	Parent Ion
					Fragment Ion(s)
normal level 30 – 150 ng/ml				pos	377
					228/198/170

detection limits	LOD	precipitation	LLE	SPE
UV	1 ng (269 nm)			
FL		5 ng/ml [2]		
ECD				
MS/MS		1 ng/ml [3]		

[1] HPLC-Fluor (453/520 nm) ACN-precipitation
[2] Analyst 110 (1985) 1505-1508, TCA-precipitation; HPLC-Fluor (328/526 nm) 5 ng/ml
[3] Clin Chem 51 (2005) 1206-1216, TCA-precipitation; HPLC-MS/MS ca. 1 ng/ml

Vitamin B6 (Pyridoxal)

• HCl

MW: 167.2

no UV
no E1/1

C_{max}-value	protein binding (PB)	elimination half life $t_{1/2}$	pKa-value	Ionisation MS	Parent Ion
					Fragment Ion(s)
40 mg oral → 200 ng/ml 5 – 24 ng/ml normal level					170
					134/152/77

detection limits	LOD	precipitation	LLE	SPE
UV	1 ng (292 nm)			
FL		2 ng/ml [1]		
ECD				
MS/MS		1 ng/ml [2]		

[1] J Pharm Sci 82 (1993) 972-974, derivatization; HPLC-Fluor (365/480 nm) 2 ng/ml
[2] Clin Chem 51 (2005) 1206-1216, TCA-precipitation; HPLC-MS/MS ca. 1 ng/ml

S
–
V

Vitamin E (Tocopherol)

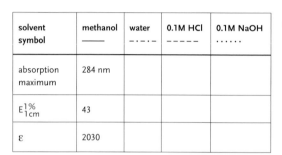

MW: 430.7

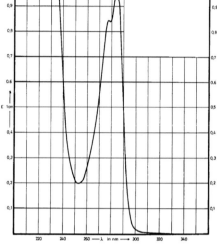

solvent symbol	methanol	water	0.1M HCl	0.1M NaOH
	———	–·–·–	– – – –	· · · · · ·
absorption maximum	284 nm			
$E_{1cm}^{1\%}$	43			
ε	2030			

C_{max}-value	protein binding (PB)	elimination half life $t_{1/2}$	pKa-value	Ionisation MS	Parent Ion / Fragment Ion(s)
5 – 20 µg/ml			12.2	pos	431 / 165

detection limits	LOD	precipitation	LLE	SPE
UV	10 ng (284 nm)		2000 ng/ml [3]	
FL			20 ng/ml [1a]	
ECD				
MS/MS			3 ng/ml [2]	

(1a) Extraction; HPLC-Fluor (292/335 nm)
(1b) Extraction; HPLC-ECD
(2) Anal Chem 79 (2007) 7087-7096, extraction; HPLC-MS/MS 3 ng/ml
(3) J Pharm Biomed Anal 35 (2004) 575-582, extraction; HPLC-UV (291 nm) 2 µg/ml

Own publication
Chronological

ppm (µg/ml)

0.5   **Allopurinol** und **Oxipurinol** HPLC-UV: Drug Res. 30 **(1980)**, 1855–1857

1   **Piracetam** HPLC-UV: J. Pharm. Biomed. Analysis 7 **(1989)**, 913–916

5   **Norfloxacin** Urin HPLC-Fluoreszenz: J. Chromatogr. A, 812 **(1998)**, 381–385

ppb (ng/ml)

100   **Benzbromaron** HPLC-UV: Drug Res. 31 **(1981)**, 510–512

5   **5-Methoxypsoralen** HPLC-UV: Drug Res. 32 **(1982)**, 1338–1341

2   **Cinnarizin** HPLC-UV: J. Chromatogr. 227 **(1982)**, 521–525

50   **Cimetidin** HPLC-UV: J. Chromatogr. 273 **(1983)**, 449–452

10   **Codein** HPLC-UV: J. Pharm. Sci. 73 **(1984)**, 1556–1558

150   **Phenytoin** HPLC-UV: Int. J. Clin. Pharmacol., Ther. and Toxicol. 22 **(1984)**, 104–107

50   **Amilorid** Harn HPLC-UV: Therapiewoche 36 **(1985)**, 56–60

2   **Amilorid** HPLC-UV: Therapiewoche 36 **(1985)**, 56–60

10   **Hydrochlorothiazid** HPLC-UV: Therapiewoche 36 **(1985)**, 56–60

75   **Indomethazin** HPLC-UV: IRCS Medical science-biochemistry 14 **(1986)**, 813–814

5   **Dextrorphan** Fluoreszenz: J. Chromatogr. 420 **(1987)**, 217–222

10   **Amanitin** HPLC-UV: J. Clin. Chem Clin. Biochem. 25 **(1987)**, 606

5   **3-Hydroxymorphinan** Fluoreszenz: J. Chromatogr. 420 **(1987)**, 217–222

1   **Nifedipin** HPLC-UV: Chromatographia 25 **(1988)**, 919–922

10   **Diclofenac** HPLC-UV: Drug Design and Delivery 4 **(1989)**, 303–311

50   **Amoxicillin** Derivatisierung + Fluoreszenz: J. Chromatogr. 506 **(1990)**, 417–421

10   **Acetylcystein** Derivatisierung + Fluoreszenz: Biopharm. Drug Disposit. 12 **(1991)**, 343–353

8   **Acyclovir** Fluoreszenz: J. Chromatogr. 583 **(1992)**, 122–127

3   **Vitamin B1** Derivatisierung + Fluoreszenz: J. Pharm. Sci. 82 **(1993)**, 56–59

2   **Vitamin B6** Derivatisierung + Fluoreszenz: J. Pharm. Sci. 82 **(1993)**, 972–974

3.5	**Silibinin** HPLC–UV: J. Liqu. Chromatogr. 16 **(1993)**, 2777–2789
1	**Triamteren** Fluoreszenz: J. Liqu. Chromatogr. 17 **(1994)**, 1577–1585
370	**Ibuprofen** Enantioselektiv Fluoreszenz: Eur. J. Clin. Pharmacol. 48 **(1995)**, 505–511
2	**Metoprolol** enant. : Int. J. Clin. Pharmacol. Ther. 34 **(1996)**, 420–423
32	**Pantoprazol** HPLC-UV : Gastroenterology 108 (Suppl.) **(1995)**, A109 oder Int. J. Clin. Pharmacol. Ther. 34 **(1996)**, 152– 156
200	**Paracetamol** ECD: Forum DR.MED **17/96**
113	**Amoxicillin** Derivatisierung + Fluoreszenz: J. Chromatogr. A 812 **(1998)**, 221–226
31	**Norfloxacin** Fluoreszenz: J. Chromatogr. A 812 **(1998)**, 381–385
30	**Minocyclin** HPLC-UV: J. Chromatogr. A 812 **(1998)**, 339–342
4.2	**Glibenclamid** Fluoreszenz: Clin. Drug Invest. 15 **(1998)**, 253–260
50	**Carbamazepin + Metabolit** HPLC-UV: Gastroenterology 110 (Suppl.) **(1996)**, A 137 oder Int. J. Clin. Pharmcol. Ther. 36 **(1998)**, 521–524
16	**Omeprazol** HPLC-UV: Methods Find. Exp. Clin. Pharmacol. 21 **(1999)**, 47–51
100	**Piroxicam** HPLC-UV: UEGW 11/**2000** Brussels
1	**Diclofenac** MS/MS: J. Pharm. Biomed. Anal. 33 **(2003)**, 745–754
40	**Carnitin + Ester** (Harn) MS/MS: Chemical Monthly 136 **(2005)**, 1425–1442
200	**Neomycin** MS/MS: J. Pharm. Biomed. Anal. 43 **(2007)**, 691–700
200	**Bacitracin** MS/MS: J. Pharm. Biomed. Anal. 43 **(2007)**, 691–700
25	**Peptid** (28mer) MS/MS: Crit Care Med 35 **(2007)**, 1730–1735
1	**Peptid** (19mer) MS/MS: LC-GC AdS, September **2007**
0,4	**Lyso-Gb3** MS/MS: Eur J Neurol 18 **(2011)**, 631–636

ppt (pg/ml)

500	**Pirenzepin**: Drug Res. 36 **(1986)**, 1409–1412
200	**Amphetamin** Derivatisierung+ Fluoreszenz: J. Liqu. Chrom. and Rel. Technol. 20 **(1997)**, 797–809
200	**Methamphetamin** Derivatisierung+ Fluoreszenz: J. Liqu. Chrom. and Rel. Technol. 20 **(1997)**, 797–809
500	**Dihydralazin** MS/MS: J. Pharm. Biomed. Anal. 43 **(2007)**, 631–645
10	**Fluticason Propionat** MS/MS: J. Chromatogr. B 869 **(2008)**, 84–92
10	**Ciclesonid + Metabolit** MS/MS: J. Chromatogr. B 869 **(2008)**; 84–92
10	**Neurotransmitters** MS/MS: SHOCK 36 **(2011)** Suppl. 1, ABS 100

ppq (fg/ml)

400	**Formoterol** MS/MS: J. Chromatogr. B 830 **(2006)**, 25–34

Index